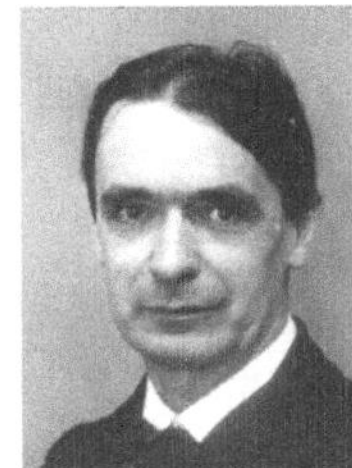
RUDOLF STEINER (1861–1925) called his spiritual philosophy 'anthroposophy', meaning 'wisdom of the human being'. As a highly developed seer, he based his work on direct knowledge and perception of spiritual dimensions. He initiated a modern and universal 'science of spirit', accessible to anyone willing to exercise clear and unprejudiced thinking.

From his spiritual investigations Steiner provided suggestions for the renewal of many activities, including education (both general and special), agriculture, medicine, economics, architecture, science, philosophy, religion and the arts. Today there are thousands of schools, clinics, farms and other organizations involved in practical work based on his principles. His many published works feature his research into the spiritual nature of the human being, the evolution of the world and humanity, and methods of personal development. Steiner wrote some 30 books and delivered over 6000 lectures across Europe. In 1924 he founded the General Anthroposophical Society, which today has branches throughout the world.

ART AS SEEN IN THE LIGHT OF MYSTERY WISDOM

*Eight lectures given in Dornach between
28 December 1914 and 4 January 1915*

RUDOLF STEINER

with an Introduction
by Marie Steiner

Translated by P. Wehrle and J. Collis

RUDOLF STEINER PRESS

Translation by P. Wehrle (lectures 1, 4–8)
and J. Collis (lectures 2 and 3)
Translation revised by J. Collis

Rudolf Steiner Press
Hillside House, The Square,
Forest Row, E. Sussex
RH18 5ES

First published under the title *Art as Seen in the Light of Mystery Wisdom* in 1984
Second revised edition 1996
Reprinted 2010

Originally published in German under the title *Kunst im Lichte der Mysterienweisheit* (volume 275 in the *Rudolf Steiner Gesamtausgabe* or Collected Works) by Rudolf Steiner Verlag, Dornach. This authorized translation published by kind permission of the Rudolf Steiner Nachlassverwaltung, Dornach

A catalogue record for this book is available from the British Library

ISBN 978 1 85584 236 6

Cover by Andrew Morgan
Typeset by DP Photosetting, Aylesbury, Bucks
Printed and bound in Great Britain by Cpod, Trowbridge, Wiltshire

Contents

Synopsis

Lecture 1, Technology and Art
Dornach, 28 December 1914

Spiritual science helps to strengthen people against the damaging influences of modern life. The influence of technology on feeling and will. Two stages of development of technology: destruction leading to expulsion of nature spirits, and creation of new connections in accordance with natural laws leading to adoption of ahrimanic laws. To withdraw from modern life is spiritual cowardice. Active participation in the ideas of spiritual science means facing up to Ahriman. Art and technology as a pendulum swing between Lucifer and Ahriman. The spirituality of language before the fifteenth century. The need for a renewal in art: the Goetheanum as a beginning. How sculpture and painting relate to music.

Lecture 2, Impulses of Transformation in the Evolution of Art—I
Dornach, 29 December 1914

How the arts relate to the being of man. The Angeloi help in sending the ego's capacity to form mental images into the astral body. Architecture: the laws of the physical body placed into external space. Sculpture: the laws of the etheric body become physical. Painting: the laws of the astral body become etheric. Music: the ego submerging in the astral body. Poetry: spirit-self lowered into the ego. Eurythmy: life-spirit lowered into spirit-self. The astral body as an example of the sevenfold structure of the different members of the human being. How the seven principles of the astral body relate to the intervals in music. The astral human being as a musical instrument. Earthly music as an imitation of heavenly music. Medieval names of the notes. Experiencing the ego through music and poetry. The ideas of spiritual science must be brought to life.

Lecture 3, Impulses of Transformation in the Evolution of Art—II
Dornach, 30 December 1914

The two aspects of spiritual science: description of cosmic matters and of a specific range of soul experiences on the path of initiation. Returning to earlier evolutionary stages of the earth through creating and enjoying art: architecture—Saturn evolution; sculpture—Sun evolution; painting—Moon evolution. Earth evolution. The two aspects of the initiation path: objective spiritual experience and subjective soul situations. Music as an external image of the initiation experience; its connection with the initiation experience; its connection with the earth's Jupiter evolution. Ancient Greeks as Sun people. The Venus of Milo shows the essence of Greek sculpture. The Goetheanum as an example of how to deal with ahrimanic influences in architecture. The boiler house as an architectural structure of ahrimanic civilization. Overcoming Ahriman by making him visible. The description of the path of initiation inspires music and poetry. The need to transform personal feeling and will into something impersonal in spiritual science.

Lecture 4, Cosmic New Year—The Dream Song of Olaf
Åsteson
Dornach, 31 December 1914

The Norwegian legend of Olaf Åsteson: by living with
the spirit of the earth during the days of Christmas the
human being as a microcosm becomes absorbed in the
macrocosm. The text of the Song. Recovering lost
ancient wisdom through the outlook of spiritual science.
Becoming immersed in the elements. The need to develop
reverence and devotion for the wisdom of ancient
revelations. A quotation from Exodus as an example.
The human being as a thought of the higher hierarchies.
Contrasts in perceptions of the sensible and spiritual
worlds. Loss of knowledge about the threefold nature of
the human being since the advent of materialism.
Applying the words liberty, equality, fraternity to different levels of human experience. Fraternity as the ideal
for the physical world, liberty for the soul world, equality
for the spiritual life. Olaf Åsteson as an image of human
beings being spiritually asleep. Dawn of a new cosmic
day.

Lecture 5, Moral Experience of Colour and Music
Dornach, 1 January 1915

Reverence on the one hand and the power of inner work
on the other arising from an understanding of spiritual
science. The position of the moon on the night of the new
year as an example of endeavouring to achieve harmony
between microcosm and macrocosm by reading the script
of the stars. The way to a new art. The experience of
colours and musical notes as an example of the moral-
spiritual experience of sense impressions. Red—divine

wrath. Shapes brought about by colours; the activity of the Elohim. Orange—longing for knowledge of the inner essence. Yellow—life in creative forces. Green—growing stronger inwardly. Blue—leading to divine mercy. Prime—being drawn into the spiritual world. Second—a sum of notes sounds towards us. Third—notes that are friends come towards us. Fourth—note-memories of different colours. Fifth—enrichment of soul experience. How spiritual forces flow into human beings. Example: a lawyer faints. Spiritual science has the task of remedying cosmic illiteracy.

Lecture 6, Working with Sculptural Architecture—I
Dornach, 2 January 1915

The dulling effect of science contrasted with the enriching and enlivening effect of spiritual science. Entering into things. Example: architecture. The balance between weighing down and supporting as a balance between the ahrimanic and the luciferic. The development of a musical mood in architecture. The Goetheanum as an attempt to reconcile the arts. The columns and architraves as the body, the domes as the soul, the windows as the spirit of the building. Enrichment of education through spiritual science. The karmic adjustment between child and teacher: the teacher's musically tuned future works on the child's sculpturally shaped past. Luciferic and ahrimanic education. Bringing spiritual science to life. The need to remain psychologically young in old age. Spiritual science as a 'draught of youth'.

Lecture 7, The Future Jupiter Evolution and its Beings
Dornach, 3 January 1915

Human beings on earth prepare future human beings for the Jupiter evolution. The relationship between science and spiritual science. Natural existence and moral action. The influence of moral and immoral actions on the beings created in the exhaled breath of human beings: preparation for Jupiter human beings or the creation of parasitic demons harking back to Old Moon. The effect of these luciferic demons on the child in the womb. Ordering life in accordance with spiritual science: allowing thought to die down in concentration and gaining equanimity in the face of destiny. Two helpful maxims. The nature of spiritual science as seen in the forms of the Goetheanum. Christian Morgenstern's deep link with the building. Seriousness and tact needed with regard to the communications of spiritual science.

Lecture 8, Working with Sculptural Architecture—II
Dornach, 4 January 1915

The architectural principle of the boiler house. Metamorphoses in the skeletal system. The double dome structure: the boiler house as an ahrimanic metamorphosis of this. The forms should be experienced inwardly rather than interpreted. Perceiving the external world by perceiving the reflections of blood and nerve activities. Pythian clairvoyance arising when the blood and nerve tracts of the body are enjoyed during waking life: reversion to Old Moon. Prophetic clairvoyance brought about by submersion in blood and nerve tracts leading to refined enjoyment of the body during sleep. New clairvoyance: taking hold of the skeletal system and turning in devotion towards the higher hierarchies. Light and dark sides of endeavours in spiritual science.

Foreword

The title *Kunst im Lichte der Mysterienweisheit* was originally given by Marie Steiner to a collection of Rudolf
Steiner's lectures brought out by her in the 1930s. The
same collection was published in English in 1935 (and in
a new translation in 1970) under the title *Art in the Light
of Mystery Wisdom*. Meanwhile, in 1966, the adminis-
trators of Rudolf Steiner's literary estate, the Rudolf
Steiner Nachlassverwaltung, had used the same German
title for the present series of lectures, which they published (together with the Introduction Marie Steiner had
originally written for the earlier collection) as Volume
275 in the Collected Works of Rudolf Steiner in German.
In 1984 the Rudolf Steiner Press, London, brought out a
translation of that volume under the present title *Art as
seen in the Light of Mystery Wisdom*.

These lectures were given in Dornach at the turn of the
year 1914–15 in the carpentry workshop located near the
first Goetheanum, then in process of being built. The
audience consisted of members of the Anthroposophical
Society living in Dornach and working on the building
during those early months of the First World War.

Introduction by Marie Steiner (abridged)

The impulses of regeneration given to mankind in these lectures will only be understood by those who are able to assimilate the nature of spiritual science fully in such a way that for them the concreteness of the spiritual world, its richness of form and being, has become a self-evident fact. Rudolf Steiner, fully equipped with the most modern scientific methods and with the utmost singleness of thought, brought the reality of the spiritual world close to his contemporaries and showed them how the human ego is placed at the focal point of the development of consciousness and how mankind must grasp this ego with full knowledge. One path towards the grasping of the ego in the fullness of life's experience, but also in sun-filled contemplation, is the path of art. It is one of the healthiest and most revealing and most direct; it was the last to turn away from its source in the temples of Mystery wisdom and has not been so quickly smothered under rubble as has the path of religion by the passion of the church for power or that of science by the rigidity of thought born of the materialistic age. For these three paths once more to find each other, for art, science and religion once more to unite and intermingle—it was for this that Rudolf Steiner worked among us. He turned his fullest attention to each of these paths because in their living synthesis he saw the salvation of mankind. They worked together once in the ancient, holy Mysteries, bringing into existence and filling with light and nourishment all the cultures of the earth. They must again be brought together and reunited through the knowledge of their undivided spiritual origin.

Human beings must now awaken within themselves a knowledge of this living and essential union. They can do this in complete freedom through careful investigation and practice if they do not timidly close themselves in the face of superior and as yet unknown powers, if they do not bow before the restraining influence of the church, nor before the authority of dogmatic science. The stages along the path have been revealed to us under the wise and entirely objective guidance of one who knows it and who, in conformity with the demands of our times, has not appealed to the human craving for submission and devotion but only to our capacity for knowledge.

The first stage is study; the basis for understanding what is presented in these lectures is the study of spiritual science. The works of Rudolf Steiner can supply the basis for penetrating the Mysteries which are the foundation of human artistic creativity and out of which impulses, acting from the supersensible sphere, must now be brought from subconscious numbness into the wakefulness of ego-consciousness. Art is beginning to wither; even art has already choked the flow from its living spiritual source. Observation through the senses, imitation of fortuitous situations on the physical plane, overemphasis of the personality: along these routes art has moved right away from its spiritual origin. To tie the torn threads together again, to recover the original spirit, once more to tread the lost road in freedom and the joy of knowledge with the newly-acquired forces of the awakened personality: this is our task today. It is hoped that the deep wisdom, the beauty and the strength speaking out of the words of Rudolf Steiner published in this book will be a help to this end.

A renewal of art will never be brought about by dallying with modern feebleness or by compromise, but only by a return to the spiritual founts of life. Having drunk at these springs, surely we cannot watch while human beings thirst without pointing out the remedy that could restore health?

The remedy lies in unlocking the wisdom of the Mysteries and presenting it to humanity in a form adapted to contemporary needs. The new initiation science must call upon our power of thought, our sense of art and style, and also the eternal essence of our being, by summoning us to conscious alertness in all these domains.

Through words, pictures and deeds Rudolf Steiner summoned us to wakefulness in each of these three domains. He has created works that give art a new orientation. He has delivered it from rigidity and brought movement into it; he has restored life to what had been smothered. Perhaps his manifold obligations in other fields would never have given him the chance to revive all the realms of art as he did, if the construction of the Goetheanum had not demanded it of him and if the World War, by curtailing some of his other activities, had not given him the time.

In the first Goetheanum, Rudolf Steiner was able to realize his living thoughts about architecture. He entrusted them to wood, the most alive of all building materials. A work of inexpressible beauty came into being, deeply affecting to the beholder by reason of the stirring force that went forth from its forms in their succession of organic developments and from the counterbalancing relations of direction—the proportions of upward and downward movement. Number, dimension and weight were triumphant in a triad of sweep, elevation and direction. The building stood as man, and man as building. The genesis of worlds, the genesis and deeds of human beings, the deeds of the gods were all inscribed in it; they were revealed in the waves of colour in the domes, in the organic growth of motifs in columns and architraves, in the luminous creation of the windows. Sculpture and painting passed beyond their own sphere, overcame line and were transformed into movement. Colour created form from within by virtue of its own creative soul quality. Music and speech became movement in the newly-blossoming art of eurythmy, and were made visible through the instrument of the human body. The creative forces of speech thus made visible were reflected back to the other forms of art, reviving them and kindling the fire of spiritual creativity. Its innate note of creativity was able to grasp the physical sound that moulds the air, filling it with spiritual substance and elevating it to higher spheres. Rudolf Steiner called his building 'the House of Speech'. All forms of art, together with science and Mystery wisdom, had found a home there. The synthesis of art, science and religion was once more accomplished.

Such a building cannot rise again, unless perhaps it shall be granted to the same individuals who executed the artistic work to transform what they have learned and experienced into another structure like it.

Otherwise a work that even by this time might have been an immense help for the spiritual development of mankind will sink back as a memory into the past, though bearing within it a spiritual seed for a new future. The flames on the eve of the New Year 1923 have for the present destroyed a mighty impulse for progress.

The retarding powers willed it so. They incited the mob against Rudolf Steiner and are still at work trying to blacken his memory. But the resounding clamour is impotent to destroy his spiritual work, for it is rooted too firmly in the soil of spiritual reality and in the needs of contemporary souls.

For the new Goetheanum the most rigid material has been employed: concrete. Like a fortress it stands, but not withdrawn in defiance. An abode for spiritual endeavour, it radiates over the landscape, welcoming all who desire to strive for that noblest of treasures, the knowledge of the world and of the human being.

The new Goetheanum does not claim to compete in form or effect in any way with its predecessor destroyed by fire. Nevertheless, its external form towers upwards with a bold and harmonious beauty, a final gift left to us by a master of the impulses of beauty. After creating its form he laid down hammer, trowel and plumb-line and worked at the word in stillness for a little while before leaving us. The work upon the building is now being carried on by his pupils according to their gifts.

It was his last bequest. The building work has been restricted by the hard constraint of inadequate funds; at the stubborn behest of financial insufficiency the artists have had to sacrifice their own intentions and often even specific directions given by Rudolf Steiner. One cannot help wondering how differently things might have turned out had there been sufficient funds to carry out also inside the new building what would be required by the force of Mystery impulses rooted in esoteric wisdom. Involuntarily again and again the thought forces its way into our minds: When and where, in what country and at what time will it be possible to erect a building that will gather together and radiate out these eternal impulses that united every detail of the burnt Goetheanum into one world-embracing whole, bringing to expression man and the universe, microcosm and macrocosm, working with spiritual creativity and soul-fashioning force?

... The following words were engraved in the windows of the first Goetheanum ... They hint at the path to be traversed through the various stages of initiation ... Genuine art leads to Mystery wisdom.

I behold
It does reveal   It has revealed

The world brings about will
Will is being born   The will has been born

The love of the world takes effect
And human love arises   And human love takes hold of him

The world gives him the power of seeing
And he sees   And he develops seeing

The outer world in resolution
Man resolving   He has willed

It has been
It had become   It was

It arises
It shall be   It is

The world wafts piety
Thus he becomes pious   Piety takes effect

The world builds
I behold the building   And the building becomes man

LECTURE 1

Technology and Art

Dornach, 28 December 1914

The main intention of lectures given here recently[1] has been to build a bridge from the knowledge of spiritual science to the kind of conception of life which our time demands, and in the coming days I intend to give a few more indications on this theme.

What we call modern life powerfully confronts those of us who have been removed from a direct connection with nature through living in towns or in similar circumstances. Since its advent people have always thought about the significance of modern life for the intellectual and material progress of human civilization, and the time has now come for the impulses we are acquiring from spiritual science to enter into this modern life. Gradually we shall have won through to the feeling that we need spiritual science as a kind of compensation for many things in modern life that weaken—we might actually say destroy—something of the general divine spiritual forces that live in us.

Those who are enabled by the first stages of the life of initiation to experience what modern civilization, in all its aspects, actually does to them will gain deeper insights into the significance it has for human existence than that obtained from an external view of life unsupported by spirituality. People who have taken the first steps in the life of initiation will pass differently through the experience of spending a night in a train or on a steamer, especially if they sleep on the journey. What is different for the person who is in these first stages of initiation, as opposed to one who has not had any connection with it, is that the experiences become conscious for the former, and he finds out what is actually happening to him when he spends a night travelling on a train or a ship, especially if he goes to sleep. Of course the person who has not acquired initiation knowledge of things also undergoes the effects that an experience of that sort has on the whole human organism. With regard to the whole effect on the human being there is, of course, no difference.

If we want to understand what these indications actually mean we must call to mind a spiritual scientific truth which you no doubt know, namely, that while we are asleep our ego and astral body are outside our physical and ether bodies.* Actually, because of certain limitations which cosmic laws impose on us in the natural order of things, our ego and astral body remain very close to our physical and ether bodies, so that if we are asleep on a train journey our ego and astral body are right inside all the rattling, rumbling and braking going on in the wheels and the engine of the train. This is equally the case on a modern steamer. We are inside everything going on around us. We are inside these not exactly musical experiences in our surroundings, and you need only have taken the very first steps in initiation to notice on waking up that when the ego returns with the astral body into the physical and ether bodies they bring

------

* Rudolf Steiner used the terms 'ether body' and 'etheric body' to denote the same principle.

with them what they experienced while they were being squeezed by the machinery, for they really were inside the moving machinery right up to the moment of waking.

We bring all this disharmonious squeezing and tearing back into our physical and ether bodies, and if you have ever woken up with the after-effects of what the engines of a steamer or a train have done to your ego and astral body, if you have brought all that into your waking consciousness, you will have noticed how little it synchronizes with what goes on within you in the way the ego and the astral body experience the inner harmony of the physical and ether bodies. You bring back with you the wildest confusion, the most frightful din of pulling, screeching and rattling, and if you are sensitive to it you will feel that the effect on the ether body really is as though your physical body were being bruised and dismembered in a machine—which is, of course, rather a drastic comparison, but you will not misunderstand it. This is an absolutely unavoidable side-effect of modern life, and I want to give a word of warning right at the outset, as the kind of lecture I shall give today can very easily arouse what I would call the veiled high-mindedness of theosophists, which flourishes all too well in some quarters.

I am not making a general allusion, of course, let alone a particular allusion, but when one gives a talk on matters like this, one immediately provokes judgements. What I mean by the high-mindedness of theosophists is that it can easily happen that people immediately imagine they must take great care not to expose their bodies to these destructive forces; that they must protect themselves from all the influences of modern life; that they must closet themselves in a room containing the right surroundings, with walls of the colour recommended by theosophy, to make sure that modern life cannot reach them in any way that would be harmful to their bodily organization.

I really don't want my lectures to have this effect. All this withdrawing and protecting oneself from the influences that we necessarily have to encounter, because of world karma, arises out of weakness. Anthroposophy can only strengthen the human being as a whole; it is intended to develop those forces that strengthen and arm us inwardly against these influences. Therefore, within our spiritual movement there can never be any kind of recommendation to cut oneself off from modern life, or to turn spiritual life into a kind of hothouse culture. This could never apply in the realm of true spiritual culture. Although it is understandable that weaker natures prefer to withdraw from modern life and go into one or another kind of settlement where they are out of reach of it, the fact remains that this arises not from strength but from weakness of soul. Our task, however, consists in strengthening our soul life by filling ourselves with the impulses of spiritual science and spiritual research so that we are armed against the onslaughts of modern life, and so that our souls can stand any amount of hammering and knocking and are still capable of finding their way into the divine spiritual realms right through the hammering and knocking of the ahrimanic spirits.

Something to which I have often referred must be taken into account here. We human beings do not only sleep at night. We actually sleep in the daytime as well, only we don't notice our daytime sleep as much as our night-time sleep. During the night our thought life is dimmed down, and because our soul lives predominantly in our thoughts we are, as a matter of course, more aware of the dimming down of our thought life during night-time sleep. During the day our life of will is more at rest, yet we are less aware of this because we live less in our will. All the arguing the philosophers have done about the freedom or lack of freedom of the will is due to this. As they have not taken into account that they are investigating the will while they are daytime sleepers and therefore cannot arrive at its true nature, they talk a lot of nonsense about free will and unfree will, indeterminism and determinism. In actual fact, while we carry on the waking life of daytime, we are only conscious of our will life to a very small degree; it is immersed in the subconscious, in the region that belongs purely to the astral body.

Thus during our waking day we are involved in all the stress and noise of technology that modern life has produced around us. During the night it is more our life of feeling and thought that becomes submerged in the noise and stress, during the day it is more our life of will and feeling.

What we call modern life has not always existed in the course of human evolution. It came on the scene essentially at the beginning of the fifth post-Atlantean era.[2] The beginning of the fifth post-Atlantean era actually coincided with the beginning of modern times. What does today's intellectual culture say about the beginning of modern times? We know that today's intellectual culture is proud of the achievements of modern life. This pride is expressed something like this: Throughout antiquity and the Middle Ages people were incapable of developing a real observation of nature such as could have led to genuine science; this did not happen until modern times. When people talk of modern times in this way they are speaking of the period which began with the fifth post-Atlantean era. That was when people broke away from the old way of observing nature and observed it impartially, solely according to its abstract laws. It was through this knowledge of the laws of nature that science came into the position of opening up the possibility of mastering the forces of nature, and mastering them in an unprecedented way, as we so often hear. This is precisely what modern technology is. The characteristic nature of modern technology arose as a result of human beings acquiring knowledge of natural laws and then proceeding to fashion matter in accordance with these laws and make machines with which they can work back on nature and life by filling modern life with them and creating their own technological setting, which is modern life in its essence and function. It is the modern age that has established real science and the resultant mastery over nature and its forces.

You often hear people speaking like this. However, if we speak like this we are speaking Ahriman's language, for this is using the language of Ahriman. Let us see if we can translate this language of Ahriman into the real and true language that we are trying to acquire anew by means of spiritual science, a language where words not only gain the meaning ascribed to them through observation of external nature, but also that ascribed to them when we look at the cosmos in its entirety, that is, both as nature and as spiritual life.

Let us begin by looking quite externally at what happens when we develop modern technology. In the first place what is happening is merely work being carried out in two stages. The first stage consists of breaking nature into pieces. We blast out quarries and take the stone away, maltreat the forests and take the wood away—and the list could go on. In short, we get our raw materials by smashing and wearing down what nature holds together. The second stage consists of taking what we have extracted from nature and putting it together again as a machine according to the laws we have come to know as the laws of nature. These are the two stages, if we look at the matter on the surface.

But what is it like if we look below the surface? Looking at it from inside, the matter is like this: When we take things from nature, mineral nature to begin with, we know from previous lectures that this is linked with a certain sense of well-being the elemental spirits have within it. This, however, is not our main concern just now. What is important here is that we cast out of nature the elemental spirits belonging to the sphere of the regular progressive hierarchies who, in fact, are the very spirits who maintain nature. There are elemental spiritual beings in all natural existence. When we plunder nature we squeeze the nature spirits out into the sphere of the spirit. This is what is constantly happening during the first stage. We smash and plunder material nature, thus extricating the nature spirits, driving them out of the sphere allotted them by the Yahveh gods into a realm where they can fly about freely and are no longer bound to their allotted dwelling places. Thus we can call the first stage the casting out of the nature spirits. The second stage is the one in which we put together what we have plundered from nature, according to our acquired knowledge of natural laws. When we construct a machine or a complex of machines out of raw materials according to our knowledge of natural laws, we put certain spiritual beings back into the things we construct.

The structure we make is by no means without its spiritual beings. In constructing it we make a habitation for other spiritual beings, but these spiritual beings that we conjure into our machines are beings belonging to the ahrimanic hierarchy. In the first stage we encounter nature spirits who are in progressive evolution and cast them out, and in the second stage we unite ahrimanic spirits with our mechanisms or other products of technology. This means that by living in this technological milieu of modern times we create an ahrimanic setting for everything that goes on in us in a sleeping state, by night or day. So it is no wonder that a person at the first stage of initiation, bringing back with him into his waking life all that he has experienced outside in the way of noise and confusion, feels its destructive character when he comes back into his physical and etheric bodies with his ego and his astral body. He is bringing back into his own organism the results of his having been in the company of the ahrimanic elemental spirits. Thus we could say that we now have a third stage, at the cultural level, a stage when in consequence of the technology all around us we are stuffing ourselves full of ahrimanic spirits. This is what things look like from the inside.

If we now turn our attention away from the occult side of modern life and look back at those times when people slept with only a thin partition dividing them from nature, a partition through which spirit could easily pass, and when their daytime work was within the realm of nature that still harboured regular spirits of the Yahveh hierarchy, we have to admit that in those times people's souls, their egos and astral bodies, brought back into their physical and ether bodies the kind of nature spirits that had an enlivening effect on their inner life of soul. The further we go back in the history of human evolution the more we find what is becoming increasingly rare today, namely, that people did not stuff themselves with the ahrimanic spirits of technology, but with nature spirits that were progressing on a straight path and which the good spirits of the hierarchies, if we may use this expression, have linked to the events and being of nature.

Human beings will only attain the kind of connection they need in order to be truly human if they seek it in their inner life, if they delve so far down into the depths of their soul that they reach the forces that connect them with the spirit of the cosmos, out of which they were born and in which they are embedded, but from which they can be separated. A separation has already taken place in their sense perception and intellect, and now again through their being filled with ahrimanic spirits in the course of modern life, as we have seen. Only by penetrating into the depths of their own being will they find the connection with divine spiritual beings that they need for their salvation, the spiritual hierarchies that are progressing along a straight path. This connection with the spiritual hierarchies from which we were actually born, in the spirit, this living connection with them, is made difficult to the highest degree by the saturation of the world by modern technology. Human beings are dragged away from their spiritual-cosmic connections, and the forces which they should be developing to maintain their link with the spiritual-soul being of the cosmos are being weakened.

A person who has already taken the first steps in initiation will therefore notice how the mechanical things of modern life penetrate into man's spiritual-soul nature to such an extent that a great deal of it is smothered and destroyed. Such a person also notices that the destructiveness of these forces makes it particularly difficult for him really to develop those inner forces which unite the human being with the 'rightful' spiritual beings of the hierarchies—please do not misunderstand the word. When someone who has taken the first steps in initiation tries to meditate in a modern railway carriage or on a modern steamer, he makes a great effort, of course, to activate the necessary forces of vision to lift him into the spiritual world, yet he notices the ahrimanic world filling him with the kind of thing that opposes this devotion to the spiritual world, and the struggle is enormous. You could call it an inner struggle experienced in the ether body, a struggle that wears you out and crushes you. Other people who have not taken the first steps in initiation also go through this struggle, of course, and the only difference is that the student of initiation experiences it consciously. Everyone has to go through it; the effects of this are experienced by everyone.

It would be the worst possible mistake to conclude from this that we should resist what technology has brought into modern life, that we should protect ourselves from Ahriman by cutting ourselves off from modern life. This would be a kind of spiritual cowardice. The real remedy is not to let the forces of the modern soul weaken and cut themselves off from modern life, but to make the forces of the soul strong so that they can stand up to modern life. A courageous approach to modern life is necessitated by world karma, and that is why true spiritual science possesses the characteristic of requiring an effort of the soul, a really hard effort.

You so often hear people saying, 'These books of modern spiritual science are difficult, they make you exert yourself in order to develop your soul forces and really penetrate into spiritual science.' This is why 'well-meaning' people—and I am saying this in inverted commas—keep on coming to me and saying that they want to smooth out difficult passages for the sake of their fellow students and change what is written in rather a difficult style into something as banal as can be—and these latter words are not said in inverted commas. However, it belongs to the essence of spiritual science that it makes demands on soul activity, that you do not accept spiritual-scientific truths lightly, as it were, for it is not just a matter of taking in what spiritual science says about one thing or another, but of how you take it in. You should take it in by dint of effort and soul activity. To make spiritual science your own you must work at it in the sweat of your soul—please forgive me for not being very polite. That is the way spiritual science operates, if you will excuse the mechanical expression.

A further misunderstanding of the actual nub of spiritual science is shown by people shying away from the difficult ideas and conceptual structures of spiritual science. Don't we know how many people shy away from it, how many people would prefer to dream—the Lord gives it to His own in sleep! They would far rather have things conjured up before them in all kinds of visions of the spiritual world than acquire knowledge through the activity of exerting their inner life of soul. We know how many people there are who prefer having visions rather than sitting down and studying a difficult book of spiritual science, even though it is capable of speaking to the human soul forces that are asleep during ordinary daily life, for spiritual science really does activate the part of us that is otherwise unconscious, transporting us into the life of the spiritual world. The right approach is not to receive conscious daily life apathetically and to grope in the dark, but to make an effort out of soul activity to get through what is given for the development of thoughts and ideas. When you make an effort and have the courage to make yourself at home in this development of thoughts and ideas, this brave and active effort will bring you to the stage where mere theorizing on what is given and mere acceptance in thought passes over into seeing and really being in the spiritual world. The really modern conception of life that arises for us from these considerations is that through our technological surroundings we descend into a kind of ahrimanic sphere and become filled with ahrimanic spirituality.

The most terrible calamity would have to come about in Earth evolution if provision had not been made in earlier ages for these experiences of ahrimanic spirituality that world karma is bringing to modern mankind. Life always progresses like the swing of a pendulum. It is experienced like a pendulum swinging in one direction and then the other. You cannot say, 'Beware of Ahriman!' for nothing can protect you from him. If someone longs to shut himself up in a room surrounded by the colour that suits him best, where there are no factories, and no trains passing by if he can possibly help it, where he is completely cut off from modern life, then there will be many, many other ways in which ahrimanic spirituality can get into his soul. Even though he withdraws from modern life, modern spirituality will still reach him.

Something has entered into human evolution that has, to some extent, warded off the calamity, and I gave an indication of this some time ago in a lecture cycle in Munich.[3] We must take all these things together, for that is also part of the active experiencing of modern spiritual science. Humanity has been given art, which also takes its raw material from nature by reducing and wearing it down, and at the second stage puts it together again to make something new, with a breath of life in it, although it is only of a pictorial nature. The life of the artistic impulses given us in the past has the capacity, as I said in Munich, to imbue its material with a more luciferic spirituality. Luciferic spirituality, beauty as an illusion, in fact everything that has an effect on us through the medium of art, leads human beings away from matter into the spirit, yet it does so through life in the material world. Lucifer is the spirit who constantly wants to flee from matter and bear us into the life of the spirit in an unauthorized way. That is the other swing of the pendulum. It is only because we have to go through a technological atmosphere in our present incarnation that it is possible for us to come into connection with Ahriman, whereas in earlier incarnations we were more connected with a quality that could be steeped in art. Thus we are countering certain luciferic forces by means of the present-day ahrimanic forces, which together form a balance; the pendulum of life, having swung one way in the past, is swinging the other way now.

What spiritual science quite specifically has to want at the present time is that human beings do not sleep and dream their way through what world karma is imposing on them. Yet people who wish to know nothing about spiritual science do sleep and dream their way through all the influences of Ahriman and Lucifer. They are exposed to these influences even though they know nothing about them. But life cannot go on like this; life has to be lived consciously from now on, and that is what spiritual science is for, so that people do not go through the world sleeping and dreaming, but understand what is around them. For this to happen, however, we must really get down to the subtleties of the way spiritual science operates—if you will forgive the expression. Such subtleties often go unnoticed, and this is the sort of thing I find when I read through transcripts of lectures I have given. Often what is of essential importance to me doesn't appear at all in the transcripts. Look at two examples of this. Just now I expressed myself in a particular way. I did not say that spiritual science wants something, but that spiritual science should want it, or has to want it. That is a particular way of putting it which comes quite naturally to a person who is speaking out of the spirit of spiritual science, for spiritual science leads as a matter of course to a more impersonal grasp of the truths of spiritual life than other sciences do. Speaking in the manner of other sciences we would say, 'Spiritual science wants something'. But spiritual science says what it 'should or must want'. And I say, 'The way I must express myself' and not, 'The way I express myself'.

A great deal depends on such subtleties; we must not pass them by. On the contrary we must begin to believe that everything depends on spiritual science taking hold of our innermost soul forces, and that it is capable of transforming them. Therefore it will not do to approach spiritual science with the kind of thinking one is in the habit of using in ordinary life. People are still largely unaware of what I mean by this. This can be seen, actually sensed, so to speak, in certain crude symptoms in the evolution of ordinary science.

Let us take one example out of many. Modern religious science—the irreligious science of religion—is especially proud of the fact that it has found concordances between New Testament utterances and commandments and Old Testament and heathen utterances and commandments. People have followed up the origin of every phrase in the Lord's Prayer, for instance, saying, 'This particular phrase comes from here and that one from there'. If you hear it like this it can sound credible. Yet the moment you approach the Mystery of Golgotha in a spiritual world-historical light you notice that all these things appear in a new context, and that the important thing is not discovering that all these expressions were there in earlier times, but looking at them in the context which gives them a new shade of meaning. In this respect the Old and the New Testament differ entirely. What has gone through the Mystery of Golgotha is linked in very subtle ways. The words and even the word connections often stay the same but their shade and colouring is different, and that makes all the difference.

There is something tremendous behind the fact, for instance, that the conception of the ego in the whole evolutionary system of language is quite differently constructed the further back we go, in pre-Christian times, than it is later on when we go forwards from the Mystery of Golgotha. The way people spoke about the 'I' changed, and this can be seen in the configuration of language. When the 'I' becomes part of the word for the verb, as is the case in many languages, it signifies something entirely different from when it is separated from the verb and spoken as a separate word, and so on.

The important thing is to work our way with the help

of spiritual science to an approach to life which looks consciously at the things that influence our human organism of spirit, soul and body. What I have described as our relationship to our technological surroundings is, of course, only in its early stages. It was about four centuries ago that things began to get like they are today. Then the nineteenth century that was so proud of itself took a tremendous leap forward in the ahrimanization of human life. Much that is important will take place in future human evolution in the direction of this ahrimanization. We have been in it for about four hundred years. It is coming slowly and gradually. It has already reached a certain climax among the vast numbers of our fellow human beings who, because of the isolation caused by living in towns, hardly have any connection any more with real nature spirits. I once said, symbolically, that it is important for a person's development to be able to distinguish oats from barley. Yet how many people are there in an urban environment today who can no longer tell the difference between oats and barley! Perhaps they can distinguish the plants, as that is comparatively easy in the case of oats and barley, but where the grains are concerned they can no longer tell one from the other. If they live in a town or were actually born there, they can usually not tell the difference.

In the evolution of mankind what happens is that when human beings have progressed a stage, this progress is always bound up with another experience that is at another stage, as it were, in a parallel stream. This is what has happened. While technological life has been drawing modern human beings closer to Ahriman in the way I have described, they have also been getting closer to him in another way. When a spiritual view of history replaces the crude way of viewing history introduced by materialism, people will understand what spiritual science must say on this matter.

In times preceding the last four centuries, human beings not only had a different relationship to their environment than they have today, but they had, above all, an entirely different relationship to something that comes to expression in them, really comes to expression in them; they had a different connection with their speech, with the way they spoke. Language does not contain only what modern materialistic science believes it contains; there is something in language which in many ways is connected with our not fully-conscious experiences, which often occur in the subconscious realms of our being, and which are therefore interpenetrated by spiritual beings. Spiritual beings live and are active in human language, and when we form words, elemental spiritual beings pour into these words. When human beings converse together spiritual beings fly about in the room on the wings of the words. This is why it is so important that we pay attention to certain subtleties of language, and do not simply let uncontrolled feelings get the better of us when we speak.

Right into the fifteenth and sixteenth centuries we could say that human beings still possessed remnants of a living experience of the elemental spirituality contained in language. The spirituality of language was still active within them, for language is in a certain respect more inspired and spiritual in many ways than an individual human being. It is only occasionally nowadays that we notice a person reverting from a materialistic way of thinking to a feeling for the inspired spirituality of language.

On one occasion here I gave a very clear if banal example showing in what way a person's mind can revert from the materialistic role of today.[4] On the whole it still happens to many people, but they are not immediately aware of it. If someone travelling down the Rhine speaks of the 'old Rhine', for instance, what does he mean? No doubt he feels something. But what is he referring to? When people speak of the 'old Rhine' I don't think they mean the river-bed, the hollow in the ground. That would be the only permanent part, of course. But we cannot discover what else the 'old Rhine' is supposed to be, for the water is certainly absolutely new; it keeps flowing on, and if you try and find anything old except the hollowed-out river-bed, it cannot be done. The old Rhine! Language is more inspired than we are, because the language obviously means the River God, even if people are not conscious of it. One is describing the elemental being that belongs to the river very suitably when one says 'the old Rhine'.

That is a rough example. This spirituality, this belief in spirituality, exists throughout language. A feeling, at least, for this connection with spirituality through language still really existed in the disposition of soul of all the peoples of Europe during the fourth post-Atlantean era and right up to modern times as far as the fifteenth and sixteenth centuries. If you are not aware of this fact you cannot have the right feeling for the beginning of the Gospel of John. The opening words of the Gospel of John, 'In the beginning was the Word', arose, in fact, out of a consciousness that the part the word plays within the whole human organism and human life is to provide a connection for human beings, initially by way of elemental spirituality, with the whole of the world lying behind the world of the senses.

If, with the means that spiritual science puts at our disposal, we observe the way human life has run its course from the Middle Ages up to modern times, and are able to look right into the soul, we shall in fact find that man's relationship to speech was altogether different in the fourth post-Atlantean era, even in the final phase of it that lasted until the fourteenth and fifteenth centuries. Whenever they spoke people heard undertones, genuine undertones. This is no longer believed, because nowadays human beings really only live in the material aspect of the sounds of speech. A spiritual element joined with the sound as though it sounded again in a lower octave. Thus when people spoke, or heard others speaking, something resounded in the words that was not differentiated according to one language or another, but was of a universal human character. One can really say that when human experience comes to expression in what is like the flowering of the separate languages, human beings today experience the flowering as a vibrating of sounds in the ear; they experience the sounds as something that has a meaning. In earlier times, however, they experienced a steeping of the whole element of speech in something that joined with it and was not differentiated into the various languages. The dividing line between the one experience and the other fell in the fifteenth and sixteenth centuries. Mankind was torn away from the geniuses of language.

Nobody can understand the real jolt mankind was given in the fifteenth, sixteenth and seventeenth centuries, unless he studies the special character of this damping down of the undertones experienced in speech.

Something was lost to human beings, and this comes to light in the happenings of the times, whether they be battles or peacetime creations. Before the point of time mentioned the human soul still experienced it. Whenever people spoke, this resounding of undertones in the experiencing of speech still lived in human souls. That is why the whole of history before this turning point has an entirely different quality from what came afterwards. Through spiritual science we must develop a spiritual ear, as I would like to call it, for the completely different sound that events had in the Middle Ages compared with that of today, since human souls were connected to their experiences in quite a different way in those times.

I can cite the Crusades as an example of a general human soul-experience. They are only conceivable in the way they came about in the Middle Ages if we know of the existence of these spiritual undertones in the experiencing of language. The present-day peoples of Middle and Western Europe would most certainly not be so affected by the words of the Council of Clermont, 'God wills it—*Dieu le veut*', as the people of the Middle Ages were.[5] The reasons for this, however, can only be recognized if we take into account what has just been said.

An important phenomenon in all modern intellectual life is also connected with this. The whole formation of modern history has to do with it. If you once envisage history with these subtle undertones of speech in mind, you will understand why, at the point of time I have indicated, the various European nationalities grouped themselves together, those nationalities who before that time had quite different relationships with one another— who were governed by quite different impulses in their relationships to one another. The way the different nationalities group themselves in the various parts of Europe, right up to the present day, has to do with impulses that we interpret quite falsely if we go back from the present to the Middle Ages to look for the origins of nations, without bearing in mind the tremendously important stage that had to be accomplished in the life of the soul.

I can only give you indications of these themes whereas they would actually require a whole series of lectures. The most important part of all this must be left to your own meditation, which will discover what can be found as a result of these indications. What I would hope to have achieved is to have given you a picture of how to build a bridge between spiritual science and knowledge of life, and shown you how spiritual science can lead to a conscious approach to the reality in which we live.

Having seen the real foundations on which these indications are based, it now appears quite obvious that this modern age of ours makes a renewal of many things necessary, compared to the past. Through being placed today by world karma in a setting that functions in an especially ahrimanic way, and through having to make our soul forces strong enough to find our way into spiritual spheres—despite all the hindrances that come to us from ahrimanic spirituality—our souls are in need of different kinds of sustenance than before. For the same reason art must also adopt new paths in all its branches.

Art obviously had to speak differently to souls that were less exposed to the attacks of Ahriman than we are today. Art has to speak in a new way to souls today, and our Goetheanum building is meant to be the very first step, really and truly the very first step towards art of this kind, and not anything perfect. It is an attempt to create the kind of art that calls on the soul to be active, along the lines of the whole conception of modern life, a spiritual conception of modern life. Let us remember the frightfully banal comparison I made regarding the Goetheanum building a few weeks ago. I asked, 'How does the effect our Goetheanum building is intended to have compare with that of an older building, or any other older work of art?'[6]

A work of art from the past made an impression by means of its forms and colours. Its forms and colours made an impression. If we make a diagram of it and show the form like this, this form had an effect on the eye [drawing]. What was in the space taken up by the form was what made the impression. And it was the same with the colours; the colours on the walls made an impression.

I said that our building was not intended to be like that; our building was meant to be—and this is the terribly banal comparison—like a jelly mould that does not exist for its own sake but for the sake of the jelly. Its function is to give a form to what is put into it, and when it is empty you can see what it is for. What it does to the jelly is the important thing. The important thing with our building is what a person who goes inside it experiences in the innermost depths of his soul, when he feels the contours of the forms. All that the forms do is set the process going that creates the work of art. The work of art is what the soul experiences when it feels the shape of the forms. The work of art is the jelly. What has been built is the jelly mould, and that is why we had to try and proceed on an entirely new principle.

Likewise the kind of paintings you will find in our Goetheanum building will not be there for their direct effect, as used to be the case with art in the past, but will be there for the soul to encounter, so that the experience resulting from this encounter will be a work of art. This of course involves a metamorphosis—I can only give indications of all this—the metamorphosis of an old artistic principle into a new one, which we can depict by saying that when the sculptural, the pictorial element is taken a stage further, it is led over into a kind of musical experience. There is also the opposite step, from the musical element back into the sculptural-pictorial.

These are things which are not created arbitrarily by the human soul, but have to do with the innermost impulses we have to go through because we are in the first third of the fifth post-Atlantean era. It has been, as it were, ordained by the spiritual beings that guide this evolution.

A start has to be made in every realm. If people find things about our building that are imperfect they may rest assured that the people who are actually engaged in building it will find far more imperfections than those who criticize it—far, far more. There are faults to be found in it which people who just look at it would not even think of. But that is not the point. The point is that a start is being made, for there are so many things that have to happen. The important thing is not the perfection we achieve in what we must will to happen, but that a start is made on what has to come to life here, however imperfect it has to be. For everything new that comes into the world is imperfect compared with old things that have stood the test of time. Things that are old have reached their highest level, whereas new creations are still in their infancy. That is quite obvious.

I will begin tomorrow where we have stopped today,

and consider the renewal of an artistic conception of the world and the connection this has with the whole cultural life of today.

LECTURE 2

Impulses of Transformation in the Evolution of Art—I

Dornach, 29 December 1914

In the course of the following considerations I shall be speaking to you about the important impulses of transformation in artistic evolution present in our time. I would like to connect this with what may occur to you as a result of your own observation of our Goetheanum building, or rather with that of which this building is merely a beginning. As a basis for these considerations we will have to begin by establishing a connection between art and the knowledge we have gained about human beings and their relationship with the world in general. I will start today with these seemingly more theoretical considerations and continue tomorrow with our actual theme concerning the impulses of transformation in artistic development.

Though I said that I would begin with a seemingly more theoretical basis, in actual fact anyone who looks upon spiritual science as something living will find these preliminaries very much alive and not at all theoretical. This will, however, only be quite clear to those for whom the ideas of physical body, ether body, astral body, ego, and so on, are not mere designations in a diagram of the human being but the expression of actual experiences in feelings and ideas relating to the spiritual world.

Of all the different forms of art, architecture appears to be the one that has become most separated from the human being as a whole. Architecture is separated from our being because it is placed at the service of our external needs, either those of utility, which call for utilitarian structures, or those having idealistic aims of many different kinds, as in the case of religious buildings. We shall see during the course of the lecture how other forms of art have a more intimate connection with our real being than has architecture. Architecture is in some way detached from what we describe as the laws of our inner being. However, seen from the point of view of spiritual science, this external character of architecture very largely disappears.

When we begin to look at the human being, the part that first strikes us, because it is the most outward, is the physical body. This physical body is permeated and penetrated and filled by the etheric body. The physical body might simply be called a space body or described as an organization in space. But the etheric body, which dwells in the physical body and, as you know, also extends beyond the limits of the physical body and is intimately connected with the whole cosmos, cannot be contemplated without the aid of time. Basically everything in the etheric body is rhythm, a cyclical rhythm of movement or activity, and it has a spatial character only in so far as it inhabits the physical body. For our imaginative perception, it is true, the etheric body also has to be conceived in spatial pictures; but these do not show its essential nature, which is cyclic, rhythmic, moving in time.

Music takes up no space, but is solely present in time. In the same way, what matters with regard to the human etheric body in reality—as opposed to the imaginative picture we draw—is mobility, movement, formative activity in rhythmic or musical sequence, in fact the quality of time. Of course, this is a difficult thing for the human mind to conceive, accustomed as it is to relating everything to space; but in order to gain a clear concept of the etheric body we must try much harder to allow musical ideas rather than spatial ideas to come to our aid.

In order to bring to the fore another characteristic of the etheric body we can say that occupying the physical body and extending, as it does, its activity and rhythmical play into this physical body, it is above all a body of forces. It is a flowing-out of forces, a manifestation of forces, and we notice these in a number of phenomena that occur during the course of our life. One of these phenomena, to which not much attention is paid by external science or from an outward view of the world, but which we have often stressed, is the ability of the human being to stand upright. On entering the world at birth we are not yet able to assume this vertical posture, which is the most important of all postures for the human being. We have to acquire the ability. It is true that this is initiated by the astral body, which somehow transfers its power of upward-stretching to the ether body, but it is the latter which, as time goes on, sets about raising the physical body into a vertical position. Here we see the living interplay of the astral and ether bodies in the formation of the physical body.

Acquisition of the upright posture is only the most striking of these phenomena. Whenever we lift a hand a similar process takes place. In our ego we can only contain the thought of lifting a hand; this thought must then immediately act upon the astral body, and the astral body transfers its activity, which lives in it as an impulse, to the ether body. And what happens then? Let us assume that someone is holding his hand in a horizontal position. Now he forms the idea, 'I want to raise my hand a little bit higher.' This inner picture, which in life is followed by the act of lifting the hand, passes over to the astral body; there an impulse arises and passes over from the astral body to the ether body. The following then happens in the ether body: the hand is at first horizontal; then the ether body is drawn up higher, then the physical hand moves, following what occurs first as a development of force in the ether body. The physical hand follows the etheric.

I shall explain the whole process tomorrow. At the moment I simply want to point out that the making of any movement involves a development of force which is followed by a state of equilibrium. The development of a force followed by a state of equilibrium is something that constantly happens in the life of our organism. Of course we have no conscious knowledge of what is really going on inside us; but what takes place is so infinitely wise that human ego cleverness is nothing by comparison. We would be quite incapable of moving a hand if we had to depend on our own cleverness and knowledge alone; for the subtle forces developed by the astral body in the ether body and then passed on to the physical body are quite inaccessible to ordinary human knowledge. The wisdom developed in this process is millions of times greater than that required by a watch-maker in making a watch.

We don't usually think of this, but this wisdom actually has to be evolved. It must be evolved and it is evolved as a result of our being left to ourselves with our ego. But the moment the ego sends the impulses of an inner picture into the astral body we need the help of another being; unaided we can do nothing. We are dependent on help from a being belonging to the hierarchy of the Angeloi. Even for the tiniest movement of a finger we need the assistance of such a being, whose wisdom is far in advance of our own. We could do nothing but lie here immobile, making mental pictures in utter rigidity, if the beings of the higher hierarchies did not constantly include us in their activity.

Therefore the first step towards initiation is to gain an understanding of how these forces act upon the human being.

I have already tried to show what is involved even in a movement as simple as resting our head in our hand. In the form of a spatial system of lines and forces we get to know the most external part of our being, which is what happens to our physical body through the activity of our ether body. If we carry this spatial system of lines and forces constantly active in us out into the world, and if we detach this system from ourselves and organize matter in accordance with it, then architecture arises. All architecture consists in separating from ourselves this system of forces and placing it outside in space. To put it another way: here we have the outer boundary of our physical body, and if we push the inner laws, which have been impressed by the ether body on to the physical body, outside this boundary, then architecture arises. All the laws present in the architectural utilization of matter are to be found also in the human body. When we project the specific organization of the human body into the space outside it, then we have architecture.

In our way of looking at things we know that the etheric body is closest to the physical body. Looking once more at any work of architecture, what can we say to ourselves about it? We can say that here, carried into the space outside us, is the interaction between vertical and horizontal and between forces that react together, all of which are otherwise to be found within the human physical body. These have been carried into the space outside us.

In the same way we can carry what streams from the astral body into the ether body, not outside ourselves this time, but down from the astral into the ether body. In other words we can bring about something which we do not separate from ourselves by placing it outside us, but which we only push further down inside ourselves. This is a process by which the laws of the ether body, which it has received from the astral body, can become physical, just as in architecture the laws of the physical body are projected into the space outside us. Through this process, sculpture arises out of the ether body, just as architecture arises out of the physical body. In a way, we push the laws of the ether body down one step.

Physical body

Architecture

Ether body

Sculpture

Just as in architecture we push the laws of the physical body into the space outside us, so in sculpture we push the laws of the ether body one step downwards. We do not separate these laws from ourselves, we push them directly into our own form. Just as we find in architecture the expression of the laws of our own physical body, so we find in sculpture the laws of our ether body; we simply transfer this inner order into our works of sculpture.

Looking at it in this way we can discover all the laws of sculpture. In architecture we transplant into the space outside ourselves the laws of the physical body, its spatial lines and interplay of forces, taking nothing of the ether body, nothing of the astral body, nothing of the ego. In the same way, where sculpture is concerned, we take only the laws of the ether body and bring them down one step lower, using nothing of the astral body and nothing of the ego except in so far as they send impulses into the ether body. This is why a work of sculpture has a lifelike appearance. It would actually be alive if it contained also the ego and the astral body. So if we seek the laws of sculpture we must realize that they are in fact the laws of our ether body, just as architecture contains the laws of our physical body.

If we do the same in connection with the astral body, in a way pushing what is in us of an astral nature a step lower down into the ether body, we are pushing further down what already lives in us. This time nothing arises that could truly have a spatial nature, for the astral body, when it moves down into the ether body, is not entering a spatial element, since the ether body is rhythmic and harmonious, not spatial. Therefore what arises can only be a picture, indeed a real picture—in fact the art of painting. Painting is the form of art which contains the laws of our astral body, just as sculpture contains the laws of our ether body and architecture those of our physical body.

Astral body

Painting

If we now take the fourth member of the human being, the ego, and push it with its laws down into the astral body, there allowing it to move and act, then we obtain yet another form of art. This art does not contain what works in the ego as something which can be expressed in language or ordinary ideas; it is something that has moved from the ego down one step towards the sub-conscious. It is as though the horizon of consciousness were to be moved down by the amount of half a member of the human being; we take half a step downward. Our ego descends into the astral body and then music is born.

Ego
Music

So music contains the laws of our ego, though not as they are manifested in ordinary, everyday life, but pressed down into the subconscious, into the astral body; it is as though the ego were to be submerged beneath the surface of the astral body, there to flow and stream within the organization of the astral body.

If we go on to speak about the higher members of the human being, starting with the spirit-self, we can refer to them only as something that is still outside the human being. For, in this fifth post-Atlantean era, we are only just beginning to make this spirit-self one of our inner members. But if we accept it as a gift from a higher sphere and sink it into our ego, if we go down into our ego, like a swimmer into water, taking with us what as yet can only be dimly felt of the spirit-self, then poetry is born.

Spirit-self
Poetry

Proceeding still further, we might say, though to a limited extent: Round about us, in the environment of soul and spirit which we shall absorb at a later stage, the life-spirit is also present. Therefore one day the life-spirit may come to be lowered into the spirit-self. But of course at the moment this is something that will only reach some degree of perfection in the very distant future, for when we try to lower the life-spirit into the spirit-self, we will have to be living entirely in an element which as yet is absolutely strange to us. So what we can say in this domain is like the babbling of an infant before it has learnt to speak properly. One can foresee for the far distant future that there will be an art of great perfection that will stand out beyond poetry, as poetry stands out beyond music, music beyond painting, painting beyond sculpture, and sculpture beyond architecture—this being of course not a question of superiority, but of arrangement. You will have guessed that I am referring to something of which we know only the most elementary beginnings today; something of which we can only receive the very first indications: the art of eurythmy. Eurythmy is indeed something that must make its appearance in human evolution at this time; but there is no call for pride, for at present it can be a mere babbling compared with what it will become in the future.

Life-spirit

Eurythmy

We can now begin at any point with a somewhat more extended view. But in order to do so we must realize that the organization of our being is not nearly as simple as we would like to imagine in our intellectual indolence.

It is incredibly easy simply to imagine that the human being consists of physical body, ether body, astral body, ego, and so on. If one is able to enumerate these various members and has an approximate conception of them, one may easily consider this rather simple form of understanding quite satisfactory. But things are not so simple. Physical body, ether body, astral body, ego are not simply shells which fit easily one into the next; they are, on the contrary, very complex structures.

Take, for instance, the astral body. It is not enough merely to say, 'This is the astral body and nothing more.' Things are much more complicated. Words are only an approximation here, but we can say that the astral body, for instance, is in itself structured and consists of seven principles. Just as the human being can be said to have seven members—physical body, ether body, astral body, ego, spirit-self, life-spirit, spirit-man—so the astral body has a connection with each of these. There is as it were a most tenuous part of the astral body which could be described as being especially moulded and fashioned for the physical body. That is to say: there is a living system of laws in the astral body for the physical body, a living system of laws in the astral body for the ether body, a living system of laws in the astral body for itself, a living system of laws in the astral body for the ego, a living system of laws in the astral body for the spirit-self, for the life-spirit, and for the spirit-man. In this way each member of the human being has in turn seven principles. So taking into consideration that the human being consists of seven members which are in turn each structured into seven principles, we already find we have a total of forty-nine.

This of course sounds perfectly horrible to modern psychologists, who like to regard the soul as a unity and would prefer to have nothing to do with these things. But for true knowledge, which must gradually arise in the course of the cultural evolution of mankind, it is certainly not without significance. When we know that the astral body is sevenfold in its nature and that it is an organism of inner living impulses, then we shall say to ourselves, 'Within this astral body, with its sevenfold organization, interactions must surely take place between its various principles.' The part of the astral body that corresponds to the physical body must have some degree of interaction with the part that corresponds to the ether body and with the part that corresponds to the astral body itself, and so on. These are no mere abstract suppositions; it is quite possible in the human organism for a person to feel inwardly—though more subconsciously than consciously—a stirring of that part of the astral body which corresponds to the physical body. Or something else may produce another stirring which will have to start in that part of the astral body that corresponds to the astral body itself, and so on. This is not a mere theory; it really happens.

Imagine the seven principles of the astral body to be interrelated in the way the notes of the scale are interrelated: prime, second, third, fourth, etc. If you allow the effect of a melody to work upon you, you will find that your human organization permits this to occur because the various notes of the melody are experienced inwardly in the corresponding principle of the astral body. The interval of a third is experienced in the part of the astral body that corresponds to the astral body itself. A fourth is experienced in the part of the astral body which—well, let us now be more specific—corresponds to the intellectual or mind soul. A fifth is experienced in the part of the astral body that corresponds to the consciousness or spiritual soul. And remembering that when we divide the human organism more accurately we find it contains nine parts, we must, accordingly, structure the astral body in the same way. Instead of saying 'the principle of the astral body that corresponds to the physical body' when enumerating the various parts, I could now say 'the principle experienced in the prime'. Instead of 'the principle of the astral body that corresponds to the ether body', I could say 'the principle experienced in the second'. And instead of saying 'the principle of the astral body that corresponds to the astral body itself' I could say 'the principle experienced in the third'.

—

—

—

Fifth	—	Consciousness soul
Fourth	—	Intellectual soul
Third	—	Astral body
Second	—	Ether body
Prime	—	Physical body

$<$ Sentient soul / Astral body

You can now also see that the existence of the major and the minor third accurately corresponds to the incorporation of the astral body in our whole human organization. If you look up the relevant passage in my book *Theosophy* you will see that there is an overlapping of what we call the astral body on the one hand with what we call the sentient soul on the other.[7] Therefore what I described as an interval of a third can correspond either to the astral body or to the sentient soul: in the one case we have the major and in the other the minor third.

It is a fact that our ability to experience a musical work of art depends upon this inner musical activity of the astral body, except that in listening to the music with our ego, we immediately bury the experience in our astral body, in certain realms that are subconscious.

This leads us to a very important fact. Let us look at ourselves as astral beings, as possessors of an astral body. What is our nature in this respect? As astral beings we have been created out of the cosmos according to musical laws. Inasmuch as we are astral beings we are musically connected with the cosmos. We are ourselves an instrument.

Let us now suppose that we do not need to hear physically sounding notes, but are able to listen to the creative activity of the cosmos that has brought us into existence as astral beings out of the cosmos. In such a state we should hear the universal music, which has always been called the music of the spheres. Let us suppose that we are able to immerse ourselves consciously in our astral entity, developing its spiritual strength to such an extent that we can hear this creative activity of the cosmic music. We could then say to ourselves, 'With the help of our astral body the cosmos is playing our own being.' This thought, which I have just expressed to you, was really alive in human beings in ancient times. In pointing this out one is also pointing to the way in which human evolution has become progressively more materialistic in the fifth post-Atlantean era. We all know that this thought does not live in the external human culture of today. People know nothing about the fact that, so far as the astral body is concerned, the human being is a musical instrument. It has not always been like this, but the fact that it was not always like this has been, so to speak, forgotten. There was a time when people said, 'A man once lived who was called John, and this John was able to transport himself into a state of spiritual consciousness in which he could hear the music of the heavenly Jerusalem.' They said, 'All earthly music can only be a copy of the heavenly music which began with the creation of mankind.' The more religious part of humanity felt that by passing into the world of physical desires, human beings had absorbed impulses which veiled and darkened the celestial music for them. But at the same time they felt that there must exist in human evolution—through purification from external and chaotic life—a road leading to the goal of hearing the spiritual cosmic music through and beyond the external music of the physical world.

In the tenth and eleventh centuries, this relation between external, materialistic music—of which the divine origin was pointed out—and its heavenly prototype, was still beautifully expressed. People were required to employ music as a form of sacrificial or religious service, and were expected to free themselves of their connection with the purely chaotic and—as it was felt to be—impure outer world when they produced musical sounds. Life in ordinary external speech was felt to be impure. People felt they were transported to spiritual heights when they raised themselves up from speech to music, which is the image of celestial music. This feeling was expressed in the following words: *Ut* queant laxis, *re*sonare fibris, *mi*ra gestorum, *fa*muli tuorum, *sol*ve polluti, *la*bii reatum, *S.J.* (Sancte Johanne).[8]

To translate this we would have to say, 'So that thy servants may sing with liberated vocal chords the wonders of thy works, pardon the sins of the lips which have become earthly'—meaning: which have become capable of speech—'O Saint John'. It was to one who could hear the heavenly Jerusalem that people looked up in this connection. Let us extract certain things that lie hidden in such a verse: *Ut* (this word was later replaced by *do*), resonare (*re*), mira (*mi*), famuli (*fa*), solve (*sol*), labii (*la*), S.J. (*si*). Thus we find that 'do, re, mi, fa, sol, la, si', the names of the notes in medieval musical notation, have been secreted into this verse.

In such an instance we can go back to what still lived in human minds right until the eleventh or twelfth centuries through atavistic clairvoyance, and we see how it all disappeared in the flood of materialistic views and passed out of people's awareness. Now, however, we are living at a time when we must find such things again and recreate them through spiritual science. Everything points clearly to the way in which evolution has made a descent so deep that a swamp has formed. The muddy water of this morass is made up of all the ideas originating in a materialistic view of the world. We are now about to struggle up again out of the mire of materialism, to ascend and rediscover what mankind has lost in its descent.

I have pointed out that properly speaking we do not only sleep by night but that certain parts of our being also sleep by day. At night it is principally the thinking and feeling parts which sleep, while during the day it is more the will and feeling parts. It is in this will and feeling sphere that we are immersed when we go with our ego into our astral body. When we hear a musical work, what happens is that we consciously go with our ego into the part of us which is otherwise asleep. When you sit and listen to a symphony, the inner process which takes place is a dulling of your ordinary, everyday thought life while you go with soul and spirit into a region that otherwise sleeps during daytime consciousness. This brings about the connection between the effect of music and all the life-giving forces in the human organism, all that streams with living force through the whole human being, letting him grow together as one with the streaming volume of sounds.

At night we sleep; then our ordinary thought life is dulled in an element of which we are as yet not aware in our normal consciousness. But if we succeed in bringing into ordinary everyday consciousness whatever it is that awakens when we sleep; if that in which we live when asleep enters into our waking experience, then ... Please note the contrast: I have just said that when we experience music the ego's awareness goes into a region which sleeps during the day. Now I am talking of immersing what we experience at night in the consciousness of daytime. When this happens, poetry arises. This is what people like Plato felt when they called poetry a 'divine dreaming'.

By exploring the connection that exists between man and the whole cosmos, which we can do to a certain extent under the guidance of art, we can bring a certain measure of life into what otherwise remains a mere scaffolding of ideas. Please do realize that these things are not a mere scaffolding! Some people take such pleasure in arranging what I have described in my book *Theosophy*[9] in the pattern of a diagram; no doubt they thought it was from pure obstinacy that I have deviated from the pattern of earlier theosophical teachings in describing three threefold organizations that are not separate but interlaced. But if these matters are approached through what one experiences and what is absolutely real, then even the nature of major and minor melodies will show that things are deeply rooted in the whole structure of the cosmos. Only when conclusions are drawn in a living way from the whole structure of the cosmos do they correspond to a true reality.

Of course it was necessary in the beginning to say a number of things the reasons for which have only appeared by degrees during the course of many years. This naturally involved the risk that people would start to criticize, because they didn't know the basis of certain statements, nor how things of necessity present themselves when the whole structure of the cosmos is taken into consideration. This is still the case with many matters. Many things that are said now are open to numerous objections if they are approached with superficial ideas, but in the course of years, perhaps decades, they will certainly be verified. The knowledge gained through spiritual science will become fruitful as soon as it is no longer a theory but a living experience.

Everything depends on our capacity to supersede the first ideas presented by the words physical body, ether body, astral body, etc., so that these ideas can become alive. A real understanding of the universe radiates from this enlivening process. Those who are able to do so should now compare the aesthetic values that have come to the fore during the last century and a half, with what can be learned from a knowledge of the human being in discovering the origin of the arts. Those who make this comparison will see that unless there is an understanding of the human organization it is impossible to reach a real comprehension of what lives around us and gives us joy.

I want to awaken in you a recognition of the fact that spiritual science is itself an initial impulse that will continue to grow and develop, that we are in a sense called upon to take the very first steps, and that we can have an inkling of what these first steps will lead to, long after we have laid aside our bodies in our present incarnation.

Physical body

    Architecture

Ether body

    Sculpture

Astral body

    Painting

Ego

    Music

Spirit-self

    Poetry

Life-spirit

    Eurythmy

LECTURE 3

Impulses of Transformation in the Evolution of Art—II

Dornach, 30 December 1914

We can gain perhaps the best survey of what ought to enter into our souls and hearts as a result of our efforts in spiritual science if we turn our attention for a moment to the greater part of what I have dealt with in my book *Occult Science: An Outline*. Leaving aside the introductory chapters, which are necessary as a preparation for the subject, we can begin with those chapters that introduce us to the being of man, his relation to birth and death, and his life in the spiritual worlds. After this comes a description of the great cosmic relationships, of course only in rough outlines; we are led through the transformations of our earth before it became the earth, through the Saturn, Sun and Moon evolutions leading up to our present Earth evolution. Then follows a sketch in the form of brief hints giving us a glimpse of the future Jupiter, Venus and Vulcan evolutions. Finally, instead of a more detailed description of the Jupiter, Venus and Vulcan evolutions we have an account of what must be undergone by the individual who wishes to set in motion within his being those inner soul experiences that must eventually lead to initiation. These processes have been described in greater detail up to a certain stage in the book *Knowledge of the Higher Worlds*.[10]

It will become apparent that for us spiritual science falls into two parts. In one part we describe cosmic relationships; we describe how what we have before us today as the earth and its beings, and as the rest of the cosmos, has been coming into existence out of the very, very distant past, and we describe the prospect of how it will develop further. If you review the many observations we have made, you will see that a large part of them are influenced by our observations concerning the development and coming into being of the cosmos.

In the other part of spiritual science we are concerned with what the soul must do in order to enter the spiritual worlds or, in other words, to reach initiation. It is these inner experiences, conquests, battles, redemptions, and achievements of the soul with which we are always concerned in this second sphere of observation. Our observations always belong essentially to one or other of these two spheres.

Beginning now with the first sphere of observations, we see that by describing the Saturn, Sun and Moon evolutions up to the present Earth evolution, we are placing something in the world which is entirely contrary to both the religious and the scientific world concepts of today; indeed the modern world for the most part considers these descriptions to be absurd. It is quite natural that to modern minds a description of a world order appropriate, for instance, to the conditions of the Saturn evolution must appear too fantastic; the description of a cosmic order of this kind must appear to people's present way of looking at things as absolute nonsense, as something that cannot exist, as the outcome of fantastic speculation. And this is just as true of the other parts of our portrayal.

Bear in mind a remark that I have made here several times including yesterday: human beings do not only sleep at night, when their conscious thoughts and ideas are dulled, but a part of their being is also asleep during the day. At night it is more the life of mental pictures that is asleep; but during the day, in a part of our being, the life of the will is more asleep. The will sleeps in the depths of our bodily being, or at least a large part of the will sleeps thus. This sphere of our will is much more comprehensive than the part that we develop consciously, which is only a small part. We may say with complete assurance that in our ordinary daily consciousness, either at work or enjoying leisure, we are really for the most part sleepwalkers. A vast number of things take place within us unconsciously; and even a great part of what seems to be done consciously is, in reality, done half or more than half unconsciously.

If we observe human beings accurately in what they do half or more than half unconsciously, then we may see with our spiritual eyes that during sleep they are not nearly so unbelieving as when awake. When awake their modern view of the world prompts them to say, 'The description of the Saturn evolution in a book like *Occult Science* is pure and absolute nonsense!' Of course they must say this. But as complete human beings they do not speak like this; for they carry within themselves something through which they—if I may say so—know unconsciously that there was once upon a time a Saturn existence. They do something which proves that in a certain way they unconsciously remember this Saturn existence: they become an architect. Architecture would never have come about if human beings did not now carry within themselves the laws that were imprinted on their physical body during the ancient Saturn evolution. Yesterday we discussed how these laws in the physical body can be projected into the space outside, where they become the laws of architecture. Human beings mysteriously project into the laws of architecture all that they took into their being during the ancient Saturn evolution. Obviously they have to use such means as are at their disposal today; accordingly the present aspect of architecture is quite different from what was implanted in us during the ancient Saturn evolution. But the essential and living elements in our architectural activities stem from what was implanted in us during the ancient Saturn evolution.

Let us enter still more deeply into the subject we are considering. What do human beings do when they become totally absorbed in the creativity of architecture, either as the architect or as the observer or admirer? They live within the Saturn laws of their physical body; if they immerse themselves entirely in the laws of architecture, they forget all about the life of their etheric body, their astral body and their ego: they become Saturn beings once more. All the impressions produced by architecture, its austerity, its chaste proportions, its silence which is yet so eloquent, result from the fact that, abandoning the higher members of our being, we immerse ourselves in what was given to us by the spirits of the higher hierarchies, the Thrones and Archai, who were active at the beginning of the Saturn evolution. It was mainly these two groups of higher spirits who were active then, assisted by the other beings belonging to the higher hierarchies.

So when we create or enjoy architecture—where it is a case of genuine art, of course—we really lift ourselves not only out of the present Earth existence but also out of the more distant past and place ourselves once more into the period of Saturn evolution.

Let us now turn to sculpture. Yesterday we saw that the laws of sculpture are the laws of the ether body which have been pressed down one step into the physical body. Just as what lives in the physical body, when compelled into the space outside, becomes architecture, so sculpture appears when what lives in the ether body is made to descend into the physical body. In enjoying sculpture we abandon the astral body, the ego, and all the higher members of our being, living as though we had only the physical body, and in the physical body an expression of the ether body. In such a condition we are once more participants of the ancient Sun evolution. All that the ancient Sun evolution implanted in us reappears when we enjoy or create works of sculpture. On the one hand these works appear so congenial to us because they give us back our own very distant past which is still creative within us, our Sun evolution; and on the other hand they are so smooth and cold in their marble because what rays out to us from them is like light coming to us from the far distances of the cosmos.

Let us now pass on to painting. We know that painting comes about when the inner impulses of the astral body are pressed down into the ether body; in painting we abandon the ego, living and experiencing as though we were only in the astral, while pressing it down into the ether body. We experience ourselves in all that the ancient Moon evolution has implanted in us, this being our inner astral nature as human beings. Painting is like the outward projection of this inner astral nature of ours. Just as in our astral nature we experience sorrow or joy, things that affect or impress us, whatever fate brings us, so do we experience what the painter conjures upon the canvas for us, which is a reflection of our own inner astral being.

If you try to enter a little into what is described in *Occult Science* as the Saturn, Sun and Moon evolutions, you will discover that an architectural mood underlies the description of the Saturn evolution, a sculptural mood that of the Sun evolution, and a pictorial mood that of the Moon evolution. The attempt was made to express these moods by choosing suitable words. The presentation of occult events definitely requires more than the current literary equipment of today. It would be an entire misconception of the style of an occult description to believe that it could be achieved by the outrageous literary devices of our time.

We now come to the Earth evolution. Here we are in the immediate present, in the reality appointed for us. What we experience here we do not immediately feel the need to place before ourselves in the form of art. But the need we feel to project our inner life outwards in the form of art is not exhausted by, as it were, recreating our cosmic past in architecture, sculpture, and painting out of the memory implanted in us.

Our need for art goes further than this and we can find the spiritual foundation for this if we turn again to the book *Occult Science*. After the descriptions of the Saturn, Sun, Moon and Earth evolutions and after a brief outline of the future Jupiter, Venus and Vulcan evolutions, we come to the description of the processes of initiation, which are essentially at first processes within the human being. These initiation processes are, in the form in which we encounter them today, the beginning of important changes for the life of human beings on earth and for the whole of future humanity.

Is it not so that our deeper experience of the life of humanity on earth is expressed in the words, 'Alas, in so far as man is consciously aware during life on earth he appears to be but an orphan in the cosmos, a child abandoned by the cosmos, or even a traveller who has lost his way in the cosmos!'

In our everyday waking consciousness we do not know the origin of what lives in us as a result of the Saturn, Sun and Moon evolutions; nor do we know what will become of us in the Jupiter, Venus and Vulcan evolutions. Knowing neither our origin nor our future we wander about at the edge of the abyss that bounds our earthly valley. Sometimes our consciousness may give us a feeling of assurance or we may feel secure as regards our future; nevertheless, neither the past nor the future can be determined objectively and competently by man on earth. But something capable of giving us clear guidance in our life will appear before our soul. This will come about when we make ourselves acquainted with the guidelines given us in the laws of initiation. Initiation in ancient times took the form of a kind of inheritance left to human beings by the gods, which manifested itself as atavistic clairvoyance; but in the course of our progress towards the future this must take hold of us more and more firmly and actually shape our inner soul life.

The path to initiation has two sides. The one side leads us to discover the secrets, the riddles of life so that we can enter into the spiritual experience of existence. The other side may be called the more subjective side of initiation that takes place more within the soul itself. It is, at the same time, the side from which people shrink the most, because it presents in fact something which does not fit in with the comfortable indolence of experience to which the soul so easily yields or wants to yield. An extremely wide and detailed range of inner experiences awaits the one who is to be gradually led by his inner experience to initiation.[11] Conquest and liberation, hindrance and redemption alternate in manifold ways with inner experience on the way to initiation. One goes through everything the soul experiences when it suddenly feels it has become an entire stranger to itself, as though it had been cast into an abyss where it cannot help feeling that it is eternally lost and can never recover anything it may have acquired during any lifetime. It may feel an unlimited dismay and grief at the loss of the existence already won. Then the soul may feel itself forced into complete fragmentation, as though it must disintegrate into an endless multiplicity and dissolve into all the beings out of which the cosmos is composed. Further, the soul may feel itself to be roaming through the beings of the cosmos, becoming akin to one of them, then leaving it again and becoming akin to another, in the way I have described in my book *The Threshold of the Spiritual World*,[12] in the part dealing with experiences that are always attended by painful privations, painful loneliness as they are passed through one by one. Then comes the experience of the most radical transformation of all, when the soul must decide to undergo what can be expressed with the words, 'Now you must lose yourself for a while, you must thrust yourself away from yourself; but you must have faith that while you are losing yourself, while you are thrusting yourself away from yourself, beings reposing in the wide expanses of the divine hierarchies will protect you, will cause you to find yourself again after you have lost yourself.' This is the passage through births and deaths. This has to be undergone among the inner experiences that lead to initiation.

At last comes the awful passage through all the forces that are not necessary for life on earth, but which are necessary for the life of the extra-terrestrial cosmos and which become the forces of evil when they are brought without justification into the life of the earth by Lucifer or Ahriman. It is the dreadful passage through the forces of evil, together with all the disruptive, devouring, engulfing forces they represent throughout the cosmos. Finally the human being passes through a stage when he ought to feel himself to be only an instrument, a tool through which the spiritual beings speak; he becomes symbolically what his larynx is as a single organ, he becomes the larynx of the divine spiritual beings, he feels himself to be resting in the all-powerful divine word. Last of all comes a condition in the future in which this feeling merges into sharing the experience of the divine will, working in the cosmos itself.

Only single stages have been described here. But the grades of experience through which the soul passes are infinite. In *Knowledge of the Higher Worlds*, and, more descriptively in *The Threshold of the Spiritual World*, you will find set forth as far as is necessary for the present time how the soul is able to adjust itself to all these states and how, at each stage, it advances one step further into the spiritual world. All that the soul passes through on the path to initiation is consciously undergone and consciously experienced.

This is why this path of knowledge is so beset with pain

and yet so full of liberation. But long, long before the human being enters consciously into all that I have described to you as the stages of the path of initiation, he is able to express these experiences in his own way in images, and this is done through music. In the final analysis genuine music is essentially the process of life taking its course in musical notes, which is an external picture of what the soul experiences consciously in the life of initiation.

If we remain in the everyday sphere we cannot at once accomplish what I described yesterday as the immersion of the ego in the astral body. To immerse the ego in the astral body in the right way is to enter into the divine world, and this is the passage through initiation. A picture of this is given to us in the processes we perceive in musical compositions. When we surrender ourselves to musical creativity, either as the composer or the listener, we abandon our ego, we push it back; but at the same time we surrender it to those divine spiritual powers who are to work upon our astral body when we have ascended to existence on Jupiter.

Please observe that we are entering upon a consideration of the creative art of music linking us with the future of mankind. It is almost, one feels, lacking in modesty to say that musical creativity is called upon to perfect itself more and more in the world, to become continually more profound, and that the musical creativity that has entered into our world so far is still more or less at the experimental stage, even though so much greatness and so much genius has already been involved. So far we have made attempts at what will be infinitely more meaningful in the musical creativity of the future. This musical art of the future will be most significantly stimulated when human beings begin to engage in getting to know the inner nature of the path of initiation.

When one day what can be described concerning the path of initiation will no longer be experienced by human beings as it is today, but will be experienced in such a way that the description of what the soul has to feel will cause human beings to go through bliss and bitter disappointment; when the knowledge that can be gained by reading about the path of initiation has become a complete inner experience, only then will it be possible for human souls to be so deeply moved through their participation in the destinies of all those beings who take part in the events of the cosmos outside the human realm that they will feel within themselves the shocks, privations and liberation that will impel the soul to express in tonal relationships what is experienced through the description of the path of initiation.

In time to come individuals will experience what is described as the path of initiation; they will feel that things which are put before us in such an apparently abstract way can be intensely experienced, much more intensely than is the case with our outward physical experience. Then a moment will come when those who are able to experience the truth of what is described as the path of initiation will say to themselves, 'Now I feel that what I am experiencing does not connect me with the realm of nature that surrounds me on the earth but with all that lives and weaves in the cosmos; not only can I experience all this, but I am able to sing it, to set it to music!'

Such a description gives us an indication of what spiritual science is to become for human beings. Spiritual science must be a living stimulus for the human soul; it should be more than mere theory, mere understanding, mere knowledge. Spiritual science must live in the soul, taking hold of all its powers, transforming the human being into another being. Or vice versa, the human being must transform himself into another being when he becomes devoted to spiritual science.

In ancient times, even in the case of the Greeks who were Sun people, an atavistic clairsentience led them to abandon their astral and their ego beings completely and only to express the laws of the physical human form created during the Saturn and Sun evolutions. Thus Greek statues were made, those works of sculpture that really stand before our physical eyes as the mankind of the Sun evolution must stand before our spiritual eyes, when we understand that the human being of that evolution consisted only of the physical human body which contained within it the living etheric forces but not as yet the astral.

A work of Greek art such as the Venus of Milo stands before us as the personification of absolute chastity, since unchastity is only possible in the astral body, in all that permeates the astral body as passion and desire. Unchastity is not possible in the etheric body. It was a heritage from the gods bestowed upon human beings that caused them to create such works of art. We have now lost this ability to feel ourselves within the ether and physical bodies alone, without the ego or the astral body.

When we awake and submerge our ego and our astral body into our ether and physical bodies, we feel and experience only what is present in our ego. Even the processes in the astral body are in the subconscious realm, and we have no inner knowledge at all of what takes place in our etheric and physical bodies. The ancient Greeks were still dimly aware of all this. But today, when we seek to bring spiritual knowledge to life once more within ourselves and cause it to embrace not only our abstract and theoretical thoughts but also the whole of our soul life, we gradually penetrate the different members of our being and learn to recognize what permeates our astral and etheric bodies in rhythmical and harmonious cycles. Then we become able to follow with our soul the ether forces that pulsate through the body and through space, calling forth forms out of the etheric.

An attempt of this kind was made in the creation of the columns and architraves in our Goetheanum building; it was a submerging into the spheres made accessible to us by spiritual science which have been forgotten by mankind. In this we must indeed take deeply seriously what spiritual science can mean to us. You can glean from all that has so far been said about spiritual science that when we enter consciously into the spiritual world—and one must enter the spiritual world consciously—in other words when with understanding we give form to what lives in the etheric world and in the human etheric body and wish to enjoy what we have thus formed, then we must inevitably make the acquaintance of those beings who are called the luciferic and ahrimanic spirits.

Consider how much of what we create is ahrimanic. You will remember what I said with regard to our modern technical environment; and of course we cannot do otherwise than use modern technology in our work.[13] If we wished to do without it we would produce the equivalent of hothouse plants. So it afforded me some satisfaction that we were able to use concrete, one of the most modern building materials, for part of our building here.[14] Progress does not consist in shutting oneself off, as in a hothouse, from the life around one, but in using what is offered by it. By grasping the spiritual nature of the world through spiritual science, we try to use modern materials in such a way that what we know through spiritual science finds a living expression in them.

Of course this is only possible up to a certain point, and you will understand why this is so when you take into account all the implications of what has been said about our technical environment. It is not possible to separate technology from ahrimanic forces if we wish, for instance, to create something for ourselves in architecture or sculpture. Thus it was a difficult task to take what of necessity had an ahrimanic nature in our building and, in a way, banish it from the building as something rendered harmless. It really was a difficult task, for we know that the ahrimanic element is inseparable from modern technology. For a while it seemed that Ahriman would easily gain the upper hand. Then we should have been compelled to incorporate into our main building all the technical equipment necessary for running it. As a result, Ahriman would have been permanently installed within the Goetheanum. We had to think of a way of excluding the ahrimanic forces from the building, and the only possibility was to take the boiler house out and make it a separate unit. This has been done, as you can see for yourselves, and with great success, a degree of success that is only possible if someone takes the trouble to enter into the requirements as understandingly as our dear Herr Englert has done.[15]

It has been possible to create, in the most modern building material, forms that truly express the following:

here, near the Goetheanum, but outside it, stands the part that must not be included in it, though it must be present outside; and the material out of which it has been formed represents an architectural structure that is truly in keeping with the knowledge of spiritual science.

It was of immense importance that this should be achieved, particularly as the most modern building materials were employed. For if you look more deeply into what I have written about spiritual science, taking in this case the last scene of *The Portal of Initiation*,[16] you will find expressed there the fact that Lucifer and Ahriman are at their most harmful when they are not seen, when they remain invisible. Suppose that somebody is tormented by ahrimanic forces. What would be the best remedy? The best remedy would be to have some kind of picture of Ahriman made and placed in that person's room. The best remedy against an astral being which torments one, is to place it before oneself in a physical form. It is incorrect to suppose that if we have Ahriman before us we will be persecuted by him; the contrary is true. Things must be made visible. But we must not let the matter get on our nerves. We must not develop a condition in which, if we happen to pass by the picture of Ahriman and look at it unconsciously, we then carry the image within ourselves, for this image will then be invisible inside us, making us nervous or agitated.

You will also see, if you study our ahrimanic chimney along with the whole boiler house, how it is indeed possible to make an architectural structure of what belongs, one might say, to the most blatant elements of ahrimanic civilization in our time. Certain defects of our civilization will not vanish until people resolve to give architectural form to the things that concern the ahrimanic elements of our civilization. Apart from all else, apart from our having a building for our own purposes, it is simply important that the first step should be taken in relating our present culture to art and in relating spiritual science to our present culture. Our boiler house is a first small step in this direction, and will lead, it is hoped, to the solution of other problems later. One enormous problem, for instance, would be to find a suitable form for the modern railway station; for the horrors and abominations which perform that function today are a contradiction of all decent humanity. In its entire form our boiler house is not only suited to its specific purpose but also corresponds to the whole relationship of Ahriman to our Goetheanum; in the same way the form of a railway station must correspond to what happens through it, with it, and in it within the framework of our modern civilization.

Such things as these should indicate the way in which spiritual science can provide inspiration for artistic creativity as well as many other fields. We may rest assured, if we enter into the true sense and spirit of what is to develop for us out of spiritual science, that one day, when human beings immerse themselves in the nature of the Saturn condition, the deeper laws of architecture will reveal themselves; and if people immerse themselves in the nature of the Sun condition, the deeper laws of sculpture will reveal themselves; and if they immerse themselves in the nature of the Moon condition, then the deeper relationships between form and colour and the nature of chiaroscuro will become apparent, creating inspiration for the art of painting.

From the description of the path of initiation will spring inspirations and intuitions for the creating of music and yet further for the creation of poetry. Then the time will come when poetic creativity in the true sense of the word will reappear in the world, for poetic creativity has to a certain extent died away. The 'divine dreaming' incorporated in the work of the true poets was the last vestige of the ancient heritage from the gods. But a time must come when, out of an understanding of the mysteries of initiation, poets will speak in dramatic or epic or lyric poetry about those intimate processes which take place in the soul when human beings are not alone with themselves, but live together with the gods of the higher hierarchies. In the not-too-distant future people will be saying, 'Stop bothering me with your perpetual jingles about men's experiences in the physical world; your daily routine of love and hate and enjoyment is your own affair.' It is about what they experience together with the gods when they have found their way outside earthly experience that poets will tell in their music and in their dramas, epics and lyrical poems. For we know that all human experiences with the extra-terrestrial world must be brought into these arts through true creativity that is not involved in everyday life.

We have now seen what impulses of transformation lie in the knowledge brought to us by spiritual science, even in the field of artistic appreciation. We have now seen how, if we enter into spiritual-scientific knowledge, we can dimly perceive the forces which must reign over the spiritual culture of future humanity. Indeed, we may be assured that without achieving a profound inner transformation nobody can really make contact with spiritual science; and we may be assured that spiritual science is something which can grasp human beings in a deeply inward way, leading beyond the narrow confines of physical life alone. If we bear in mind this ideal of spiritual science, if we bear in mind that spiritual science can lead out into a sphere that is different from ordinary experience, then it is always an event of immense significance to see somebody within the movement of spiritual science really igniting within him or herself the spark that leads out of and beyond the narrow limits of ordinary personal experience. In a way, the only joyful experience which is so far made possible for us through spiritual science is that, as a result of this movement, individuals can appear among us who really find their way out of their personal sphere into those spheres where the personal element ceases to exist.

In our everyday life we must, of course, cultivate the personal element; but in so far as we are together as students of spiritual science, all personal will and feeling is changed into something impersonal if we take hold of spiritual science in the right way. Every victory over personal feelings and over the weight of personal matters in life is of immense significance and value. But on the other hand it is one of the bitterest disappointments if something that is striven for in spiritual science, with a will that is purely spiritual, becomes entangled again in the merely personal will and purpose of human beings, and if personal matters begin to play a role within a society whose object it is to unite us in striving for the knowledge of spiritual science.

I shall not elaborate on these concluding remarks or go into more detail because I believe that there are a good many among you who will perhaps understand much of what is meant by them, who will understand that they were intended as an indication of a number of things that are satisfactory and a number that are disappointing.

Today, having tried for a while to walk together along a path of spiritual science, it is good to think about these things for a moment; for there are various reasons why we should reflect and ask ourselves to what extent our own soul is participating in the sincere and honest effort to achieve the spiritual purposes which are nourished by the current of spiritual science. What a superb perspective unfolds before us when we say, 'Life, science, religion and also art can receive impulses of transformation from spiritual science when it is truly understood. In the case of the pictorial and sculptural arts the impulses come from what we learn in spiritual spheres about the past; and in the case of the arts involving music and speech the impulses come from what we are striving for inwardly, in order to be able to approach the future.' This perspective is so immense and so powerful that we cannot bring enough realization to bear on it, in order to make it more intensely clear to ourselves. The more we succeed in making clear to ourselves the inner mood resulting from this vision, the better shall we be as true members of that great organism known to us as spiritual science, an organism which is small today, but which has great possibilities within it. Not only do I want to appeal to your reason and your understanding with this; I also want to sow it as a seed in your souls and in your hearts.

LECTURE 4

Cosmic New Year—The Dream Song of Olaf Åsteson

Dornach, 31 December 1914

To begin our end-of-year festival Frau Dr Steiner will recite for us the beautiful Norwegian legend of Olaf Åsteson, of whom we are told that at the approach to Christmas he fell into a kind of sleep which lasted for 13 days: the 13 holy days that we have explored in various ways. In the course of this sleep he had significant experiences which he was able to recount when he awoke.

During the last few days we have examined various things that make us aware of how spiritual science gives us a new approach to an understanding of gems of wisdom which, in past times, people realized belonged to spiritual worlds. Time and again we shall encounter this prehistoric knowledge of the spiritual worlds in one instance or another, and we shall continually be reminded that what was known in former ages was due to the fact that human beings were so organized then that they had the kind of relationship with the whole of the cosmos and its happenings that we would now call being immersed with their human microcosm in the laws or activities of the macrocosm, and that in this process of immersion in the macrocosm they were able to experience things that deeply concern the life of the soul, but which are hidden from us as long as we live as microcosm on the physical plane and are equipped only with a knowledge given us by our senses and an intellect bound to those senses.

We know that only a materialistic outlook can believe that human beings alone in the world order are equipped with thinking, feeling and will, whereas a spiritual point of view must acknowledge that just as there are beings below the human level, there are also beings above the human stage of thinking, feeling and will. Human beings can live in their own way into these beings when, as microcosm, they immerse themselves in the macrocosm. However, we then have to speak of the macrocosm not only as a macrocosm of space but also as one in which the course of time is of significance. Just as human beings, in order to kindle the light of the spirit within them when they want to descend into the depths of their own soul, have to shut themselves off from all the impressions their environment can make on their senses and have, as it were, to create darkness round them by closing off their sense perceptions, likewise does the spirit, whom we can call the spirit of the earth, have to be shut off from the impressions of the rest of the cosmos. The outer cosmos has to have least effect on the earth spirit if the earth spirit is to be able to concentrate its forces and capabilities within. For then the secrets will be discovered that human beings have to discover in conjunction with the earth spirit, as a result of the earth having been separated as Earth from the cosmos.

The time when the outer macrocosm exercises the greatest effect on the earth is the time of the summer solstice, midsummer. Many accounts of olden times connected with festive presentations and rituals remind us that festivals like these take place at the height of summer; that in the midst of summer, the soul, in letting go the ego and merging with the life of the macrocosm, surrenders in a state of intoxication to the impressions from the macrocosm.

Conversely, the legendary or other kind of presentations of what could be experienced in olden times remind us that when impressions from the macrocosm have least effect on the earth, the earth spirit, concentrated within itself, experiences within the eternal All the secrets of the earth's life of soul, and that if human beings enter into this experience at the point of time when the macrocosm sends least light and warmth to the earth, they learn the holiest of secrets. This is why the days around Christmas were always kept so sacred, because while his organism was still capable of sharing in the experience of the earth, man could meet the spirit of the earth during the point of time when it was most concentrated.

Olaf Åsteson, Olaf the son of earth, experiences various secrets of the cosmic All while he is transported into the macrocosm during these 13 shortest days. The Nordic legend, which has recently been extricated from old accounts, tells of these experiences Olaf Åsteson had between Christmas and New Year up until the sixth of January. We often have reason to remember this former manner in which the microcosm took part in the macrocosm, and we can then take these things further. First of all, however, let us hear the legend of Olaf Åsteson, the earth son who, during the season in which we now are, experienced the secrets of cosmic existence in his meeting with the earth spirit. Let us listen to these experiences.[17]

THE DREAM SONG

I
Come listen to my song!
The song of a nimble youth.
    Of Olaf Åsteson will I sing,
    Who lay and slept so long.

II
He laid him down on Christmas Eve
And soon lay deeply sleeping.
Nor could he awaken
Until the people went to church
Upon the thirteenth day.
    Of Olaf Åsteson will I sing,
    Who lay and slept so long.

He laid him down on Christmas Eve
And he slept long indeed!
He could not awaken
Until the bird was on the wing
Upon the thirteenth day.
    Of Olaf Åsteson will I sing,
    Who lay and slept so long.

Olaf could not awaken
Until the sun shone o'er the peaks
Upon the thirteenth day.
Then saddled he his nimble horse
And rode in haste to the church.
    Of Olaf Åsteson will I sing,
    Who lay and slept so long.

The priest was at the altar
Reading holy mass
When Olaf alighted at the gate
To tell the many dreams
That had passed through his soul
When he did sleep so long.
        Of Olaf Åsteson will I sing,
        Who lay and slept so long.

Then old and young they all gave heed,
To Olaf's words they harkened
That told them of his dreams.
        Of Olaf Åsteson will I sing,
        Who lay and slept so long.

III

'I laid me down on Christmas Eve
And soon lay deeply sleeping.
Nor could I waken
Before the people went to church
Upon the thirteenth day.
        The moon shone bright
        And all the paths led far away.

I was borne up into the clouds,
Thrown down to the ocean's depths,
And whosoever will follow after
Good cheer he will not find.
        The moon shone bright
        And all the paths led far away.

I was borne up into the clouds,
Then hurled into murky swamps,

And I saw the horrors of hell
And also heaven's light.
    The moon shone bright
    And all the paths led far away.

I had to go through deep, dark clefts
Where heaven's rivers rushed and roared.
The power to see them was not mine
Yet I could hear their roaring.
    The moon shone bright
    And all the paths led far away.

My coal-black horse he did not neigh,
Nor did my good hounds bark,
The bird of morning did not sing
For a wonder lay on all.
    The moon shone bright
    And all the paths led far away.

I had to travel in spiritland
Through stretch on stretch of thorny heath,
My scarlet mantle was torn to shreds
The nails of my feet likewise.
    The moon shone bright
    And all the paths led far away.

Then I came to the Gjallar Bridge[18]
Suspended in the windblown heights,
Studded it is with rich red gold
And the nails thereon have sharp points.
    The moon shone bright
    And all the paths led far away.

The spirit snake he struck at me[19]
The spirit hound bit me,
And lo! the bull did bar my way.
These are the three beasts of the bridge,
Most wicked are they all.
　　　The moon shone bright
　　　And all the paths led far away.

The hound he is a snappish beast
The serpent waits to strike,
The bull is ready to attack!
And no one may pass o'er the bridge
Who will not honour truth!
　　　The moon shone bright
　　　And all the paths led far away.

I passed o'er the Gjallar Bridge
On dizzy heights and narrow.
I who had waded in the swamps...
Behind me now they lie!
　　　The moon shone bright
　　　And all the paths led far away.

I had waded in the swamps,
There seemed no foothold I could find.
As I passed o'er the Gjallar Bridge
Earth did I feel within my mouth
As the dead who lie in their graves.
　　　The moon shone bright
　　　And all the paths led far away.

To the waters then I came,
'Twas where the icy masses gleamed

Like unto flames of blue...
And God did guide me in my steps
That I did not come close.
        The moon shone bright
        And all the paths led far away.

So I went on the wintry way
And saw on my right hand:
Like unto paradise it was,
Light shining far and wide.
        The moon shone bright
        And all the paths led far away.

God's Holy Mother then I saw
Amidst most wondrous glory!
"Now take thy way to Brooksvalin,[20]
The place where souls are judged!"
        The moon shone bright
        And all the paths led far away.

IV

In other worlds I tarried then
Through many nights and long;
And God alone can know
The suffering I saw there,
        In Brooksvalin, where souls
        World judgment undergo.

I could see a young man
Who in life had killed a child.
Now he must carry it always
And stand in mud to his knee,
        In Brooksvalin, where souls
        World judgment undergo.

Also I saw an old man
Wearing a cloak of lead;
Thus was he punished,
The miser on earth,
    In Brooksvalin, where souls
    World judgment undergo.

And men appeared before me
Wearing apparel of fire;
So does their dishonesty
Weigh on their poor souls,
    In Brooksvalin, where souls
    World judgment undergo.

Children I also saw,
Glowing coals beneath their feet,
In life they did their parents ill,
Now must their spirits feel it,
    In Brooksvalin, where souls
    World judgment undergo.

And to a house I had to go
Where witches toiled in blood;
This was the blood of those
Who had enraged them while on earth,
    In Brooksvalin, where souls
    World judgment undergo.

Now there came riding from the North
Wild hordes of evil spooks,
Led by the Prince of Hell,[21]
    In Brooksvalin, where souls
    World judgment undergo.

This horde riding from the North
Was the wickedest ever seen;
And the Prince of Hell rode out in front,
And he rode on his coal-black steed,
>   In Brooksvalin, where souls
>   World judgment undergo.

Yet now came a host from the South
Bringing holy calm,
And at their head rode Saint Michael
At the side of Jesus Christ,
>   In Brooksvalin, where souls
>   World judgment undergo.

The souls weighed down by sin
Had to tremble in anguish and fear!
Their tears ran down in streams
To hear of their wicked deeds,
>   In Brooksvalin, where souls
>   World judgment undergo.

Michael stood in majesty[22]
And weighed the souls of men
Upon his heavenly scales.
And near him, judging, stood
The Lord of Judgment, Jesus Christ,
>   In Brooksvalin, where souls
>   World judgment undergo.

V

Blessed is he who in earthly life
Unto the poor gives shoes!
He does not need, with naked feet,

To walk on the heath of thorns.
        Thus speaks the Balance,
        And World truth
        Sounds forth in spirit heights.

Blessed is he who in earthly life
Unto the poor gave bread!
For nothing of harm can come to him
From the hounds of spiritland.
        Thus speaks the Balance,
        And World truth
        Sounds forth in spirit heights.

Blessed is he who in earthly life
Unto the poor gave corn!
The horns of the bull are no threat to him
When he crosses the Gjallar Bridge.
        Thus speaks the Balance,
        And World truth
        Sounds forth in spirit heights.

Blessed is he who in earthly life
Unto the poor gives clothes!
He need not fear the freezing wastes
Of ice in Brooksvalin.
        Thus speaks the Balance,
        And World truth
        Sounds forth in spirit heights.'

VI

And young and old they all gave heed,
To Olaf's words they harkened

That told them of his dreams.
You have slept long indeed...
Awaken now, O Olaf Åsteson!

My dear friends, we have just heard how Olaf Åsteson fell into a sleep that was to reveal to him the secrets of worlds that are hidden from the world of the senses and ordinary life on the physical plane. This legend brings us tidings of ancient knowledge and insight into the spiritual world, which we shall regain once more through what we call the world view of spiritual science.

You have often heard it said in connection with the human soul's entry into the spiritual world that human beings behold the spiritual world only when they experience the gates of death and then enter into the elements. This means that the elements of earth existence do not then surround them in the way they do in ordinary life on the physical plane, in the form of earth, water, air and fire, but that they are lifted above this sensory exterior of the elements and enter into what these elements really are when you know their true nature, where beings exist that have a relationship with human soul experience.

We were able to sense that Olaf Åsteson experienced something of this descent into the elements when we came to the part where Olaf reaches the Gjallar Bridge and crosses over it on to the paths of the spiritual world that all lead far away. What a vivid description we are given of his experience as he descends into the element of earth. It is described in such detail that he tells us he himself feels earth in his mouth like the dead who lie in their graves. Then there is a clear indication of his going through the element of water, and of all that can be experienced in the watery element when one also experiences its moral quality. He also indicates how the human being meets with the elements of fire and of air.

All this is described in a wonderfully graphic way and centred in the experience of the human soul meeting the secrets of the spiritual world. The legend was found at a later date; it was collected at the place where it lived among the people as an oral tradition. Parts of the legend in their present form are no longer the same as in the original. Doubtless the graphic description of the experiences in the earth realm originally came first and then the experiences in the realm of water. The experiences in the realms of air and of fire were no doubt far more differentiated than they are in the feeble after-echo that we have today, which was found centuries later.

The conclusion was undoubtedly also much more impressive and less sentimental, for in its present form it does not in the least remind us of the sublime language of olden times, nor of the capacity to raise the listener on to a superhuman plane that used to exist in folk legends. The present conclusion lives merely on a human level, and the reason why it is moving is purely because of its connection with such deep secrets of the macrocosm and of human experience.

If we rightly understand the season of the year in which we now are, we have a strong urge to remember the fact that humanity used to possess a knowledge— even if it was less defined and clear-cut—that has been lost and which has to be regained. As we certainly recognize today that that particular kind of knowledge has to return if mankind is to be made whole, we can surely ask ourselves whether we should not consider it one of our most urgent tasks to do everything we can to bring knowledge like this into the culture of the present.

Many things will have to happen in order for this change to come about in the right way, in what I would like to call the feeling content of our world conception. One thing will be particularly necessary—I say one, for although it is one among many, you can only take one at a time—and this is that it will be essential for human souls, on the basis of spiritual science and the world view it represents, to acquire reverence and devotion for what was known in ancient times in the old manner about the deep secrets of existence. People must arrive at the feeling that during the materialistic age they have neglected the development of this reverence and devotion.

We must get the feeling of how dried-out and empty this materialistic age is, and how proud of their intellectual knowledge people were in the early centuries of the fifth post-Atlantean era in face of the revelations of ancient religion and knowledge handed down from former times, which, when approached with the necessary reverence, truly give us the feeling that they contain the most profound wisdom. Fundamentally speaking we have no reverence for the Bible nowadays, either! Disregarding the kind of atrocious modern research that tears the whole Bible to shreds, we have merely to look at the dry and empty way we approach the Bible today armed, as it were, only with the knowledge of the senses and ordinary intellectual powers, and at the way we can no longer muster a feeling for the tremendous greatness of human perception that comes to meet us in some of its passages. I would like to refer to a passage from Exodus, Chapter 33, Verse 18:

And Moses said to God, 'I beseech thee, shew me thy glory.'

And Yahveh said, 'I will make all my goodness pass before thee, and I will proclaim the name of Yahveh before thee; and will be gracious to whom I will be gracious, and will shew mercy on whom I will shew mercy.'

But then Yahveh said, 'Thou canst not see my face: for there shall no man see me, and live.'

And Yahveh said, 'Behold, there is a place by me, and thou shalt stand upon a rock:

And it shall come to pass, while my glory passeth by, that I shall put thee in a cleft of the rock, and will cover thee with my hand while I pass by:

And I will take away mine hand, and thou shalt see my back parts: but my face shall not be seen.'

If you gather together various things we have taken up in our hearts and souls during the years we have been working with spiritual science, and then approach this passage, you can have the feeling that infinite wisdom is speaking to us there and how, in the materialistic age, human ears are so deaf that they hear nothing of the infinitely profound wisdom that comes to us from this passage. I would like to take this opportunity to refer you to a booklet that has been published under the title *Worte Mosis*, Bruns Publishing Co. in Minden, Westphalia, because certain passages from the five Books of Moses have been translated [into German] better in this booklet than in other editions. Dr Hugo Bergmann, the publisher of *Worte Mosis*, has taken a lot of trouble over the interpretation.[23]

We have often stressed the fact that if we want to

penetrate to the spiritual world we will have to acquire a totally different relation to that world than the one we have to the sense world. We have the sense world all around us, we look at it and see it in its colours and forms and hear its sounds. The sense world is there, and we are in the midst of it, feeling its influence, perceiving it and thinking about it. That is how we relate to the sense world. We are passive and in a manner of speaking the sense world works its way into our souls. We think about the sense world and make mental images of it.

Our attitude is quite different when we penetrate into the spiritual world. This is why it is so difficult to get the right idea of what a person experiences when he enters the spiritual world. I have attempted to characterize some of these difficulties in my small book *The Threshold of the Spiritual World*.[24] We make mental images of the sense world and we think about it. If we go through everything a person has to go through if he wants to follow the path of initiation, something occurs that can be described like this: we ourselves relate to the beings of the higher hierarchies in the same way as the things around us relate to us; the higher beings make a mental image of us, they think us. We think the objects around us, the minerals, plants and animals; they become our thoughts. We in turn are the conceptions, thoughts and perceptions of the spirits of the higher hierarchies. We become the thoughts of the Angeloi, Archangeloi, Archai and so on. They take us in, in the same way as we take in plants, animals and human beings. We should feel sheltered and protected when we say, 'The beings of the higher hierarchies think us, they make mental images of us. These beings of the higher hierarchies take hold of us with their souls.' In fact we can actually picture that when Olaf Åsteson fell asleep he became a mental image of the spirits of the higher hierarchies, and in the course of his sleep these beings of the higher hierarchies experienced what the beings of the earth spirit experience (these are, of course, a plurality for us). When Olaf Åsteson then sinks back down into the physical world he remembers what the spirits of the higher hierarchies experienced in him.

Let us imagine for a moment that we are setting out on the path of initiation. How can we relate to the spiritual world, which is a host of spiritual beings of the higher hierarchies, into which we wish to enter? How can we relate to them? We can appeal to them and say, 'How can we enter into you, how do you reveal yourselves to us?' Then, when we have acquired an understanding of the different kind of relationship the human soul has to the higher worlds, there will sound forth to us from the spiritual worlds:

'You cannot perceive the spiritual world in the same way as you perceive the sense world, in the way the sense world appears before you and impinges on your senses. We must think you, and you must feel yourself in us. You must feel the kind of experience in you which a thought you think in the sense world would have if it could experience itself within you. You must surrender yourself to the spiritual world, then the beings of the higher hierarchies who can reveal them-selves to you will enter into you. This will stream into your soul and live within it, bringing grace, in the same way as you live in your thoughts when you think about the sense world. If the spiritual world wishes to favour you and be compassionate towards you, it will fill you with its love! You must not imagine that you can approach spiritual beings in the same way as you approach the sense world. Just as Moses had to creep into the cave, the "cleft of the rock", you must go into the cave of the spiritual world. You have to put yourself there. Like a thought lives in you, you must be taken up into the life of the spiritual beings. You yourself must live as a universal thought in the macrocosm. To have experiences there of your own accord is not possible during earthly life between birth and death, but only after you have passed through death. No one can experience the spiritual world in this way before he has died, yet the spiritual world can come close to you, bless you and fill you with its love. And if after, or while, you are within the spiritual world, you develop your earthly consciousness, the spiritual world will shine into this consciousness.'

As we confront an object when it is outside us, and when it enters our consciousness it is inside us, so is the soul of man within the cave of the spiritual world. The spiritual world passes through him. Here, the human being confronts things. When he enters the spiritual world the beings of the higher hierarchies are behind him. There, he cannot see their face, just as a thought cannot see our face when it is within us. Our face is in front and the thoughts are behind, so they cannot see our face. The whole secret of initiation is concealed in the words Yahveh speaks to Moses.

And Moses said to God, 'I beseech thee, shew me thy glory.'

And Yahveh said, 'I will make all my goodness pass before thee, and I will proclaim the name of Yahveh before thee; and will be gracious to whom I will be gracious, and will shew mercy on whom I will shew mercy.'

But then Yahveh said, 'Thou canst not see my face; for there shall no man see me, and live.' (Initiation does indeed bring you to the Gate of Death.)

And Yahveh said, 'Behold, there is a place by me, and thou shalt stand upon a rock:

And it shall come to pass, while my glory passeth by, that I shall put thee in a cleft of the rock, and will cover thee with my hand while I pass by:

And I will take away mine hand, and thou shalt see my back parts; but my face shall not be seen.'

It is the opposite of the way we perceive the sense world. You have to muster a lot of the effort you have developed over the years in spiritual science, in order to encounter a revelation like this with the right kind of reverence and devotion. Then human souls will gradually acquire more and more of this feeling of reverence towards these revelations; and this reverence, this devotion, is among the many things we need in order that the change we have been speaking of can come about in mankind's spiritual culture.

The time when the macrocosm sends down least influence to the earth, the days from Christmas over New Year until roughly the sixth of January, can be a suitable time not only for remembering the facts of spiritual knowledge, but also for remembering the feelings we have to develop as we take up spiritual science. We are really and truly taken up again into the life of the spirit of the earth, together with whom we form a whole, and in which ancient clairvoyant knowledge lived, as this legend of Olaf Åsteson shows us. Humanity in the materialistic age has in many ways lost this reverence and devotion for spiritual life. It is most essential to see to it that this reverence and devotion comes back, for without them we shall not develop the mood to approach the new spiritual science in the right way. Unfortunately the mood with which spiritual science is approached is still the same mood we have for ordinary science. A thorough change will have to come about in this respect.

Having lost the understanding for the spiritual world, human beings have also lost their proper relation to humanity as a whole. The materialistic world conception produces chaotic feelings about universal existence. These chaotic feelings about the world and humanity were bound to come in the age of materialism. Think of a time—and this is our time, the first centuries of the fifth post-Atlantean cultural era—when people no longer have any real awareness that the being of man is three-fold: a bodily nature, soul and spirit. For it really is like this. The threefold nature of the human being, which to us is one of the basic elements of spiritual science, was something that people have not had the slightest notion of from the first four centuries of the fifth post-Atlantean cultural era right into our time. Man was just man, and any talk of structuring his being in the way we do into body, soul and spirit was considered complete nonsense.

You might imagine that these things only have meaning in the sphere of knowledge, but that is not so. They are important not only as knowledge, but for the whole manner in which human beings face life. In the fourth century of modern times, or, as we say in our language, during the fifth post-Atlantean cultural era, three great words came to the fore in which people saw, or at least endeavoured to see, the essence of human endeavour on earth. Important though these words are, what made them significant was the fact that they appeared at a time when mankind knew nothing of the threefold nature of man. Everyone heard of liberty, equality and fraternity.

It was a profound necessity that these words were heard at a certain time in modern civilization, but people will only really understand them when the threefold structure of the human being is understood, because until then they will not realize the significance these words can have with regard to man's real being. So long as these words are approached with the sort of chaotic feelings that are engendered by the thought that man is man, and the threefold structuring of the human being is nonsense, people will find no guidance in these three words. For the three words, as they stand, cannot all be directly applied to one and the same level of human experience. They cannot be. Simple considerations which do not perhaps occur to you because they seem too simple for such weighty matters, can go to show that if they are taken on the same level, what these three words mean can lead to serious conflict.

Let us start by looking at the realm where we find fraternity in its most natural form. Take human consanguinity, the family, where there is no need to instil brotherly love because it is inborn, and just think how it warms the heart to see real genuine brotherhood among a family, to see everyone united in a brotherly way. And yet—without losing any of the wonderful feeling we can have about this brotherly love—let us have a look at what can happen to a family fraternity precisely because of this brotherliness. Brotherliness is justified within a family, yet a member of a family can be made unhappy by it, and can long to get away from it because he feels he cannot develop his own soul within the family fraternity and must leave it in order to develop in freedom. In this way freedom, the unfolding in freedom of the life of the soul, can come into conflict with even the best-meant brotherliness.

Obviously a superficial person might maintain that it is not proper brotherliness if it does not agree with a person's freedom. But people can say anything they like. No doubt they can say that everything agrees with everything else. I recently saw a doctoral thesis in which one of the articles that had to be proved was that a triangle is a quadrangle. You can of course plead for a thing like that, you can even prove exactly that a triangle is a quadrangle! And you can also fully prove that fraternity and freedom are compatible. But that is not the point. The point is that for the sake of freedom many a realm of brotherliness has to be—and in fact is—forsaken. We could give further examples of this.

If we wanted to count up the discrepancies between fraternity and equality it would take us a long time. Obviously we can say *in abstracto* that everyone can be equal, and can show that fraternity and equality are compatible. But if we take life seriously it is not a question of abstractions but of looking at reality. The moment we realize that the human being has a bodily nature that lives on the physical plane, a soul nature that actually lives in the soul world, and a spiritual nature that lives in the spiritual world, we have the right perspective for the connection between these three profound words. Brotherliness is the most important ideal for the physical world, while freedom is for the soul world. In so far as man enters into the realm of the soul we ought to speak of the freedom of the soul, that is, of the kind of social conditions that fully guarantee the soul its freedom. Then, if we bear in mind that in order to develop the spirit and enter the spirit land we, that is, each one of us, have to strive for spirit knowledge from our own point of view, we shall soon see where we would get with our spiritual conceptions if each one of us only went his own way and we all filled ourselves with a different content.

As human beings we can only find one another in life if we seek the spirit, each one for himself, yet can arrive at the same spiritual content. We can speak of the equality of spiritual life. We can speak of fraternity on the physical plane and with regard to everything that has to do with the laws of the physical plane and which affects the human soul from the physical plane; liberty with regard to all that comes to expression in the soul in the way of laws of the soul world; equality with regard to everything that comes to expression in the soul in the way of the laws of the spirit land.

So you see, a Cosmic New Year must come about, where there will be a sun that will increase in power to give warmth and to radiate light: a sun that must bring light-filled warmth to many a thing that lived on during the age of darkness, yet was not understood. It is characteristic of our time that a good deal is striven for and expressed in words, yet is not understood.

This, too, can bring us to feel reverence and devotion for the spiritual world. For if we ponder on the fact that many people strove for fraternity, liberty and equality in the fourth century of the fifth post-Atlantean era and uttered these words without understanding them properly, it is possible for us now to see an answer to the question, 'Where did these words come from?' The divine-spiritual universal order implanted them into the human soul at a time when we did not understand them, in order that key words of this kind might lead us on to true universal understanding. We can notice the wise guidance in world evolution even in things like this. We can observe this guidance everywhere, whether in past ages or in more recent times, observing that often we do not notice until afterwards that something we did previously was actually wiser than the wisdom we had at our command at the time. I drew attention to this at the very beginning of my book *The Spiritual Guidance of Man*.[25]

However, if you look at the fact that in world evolution, in the evolution of mankind, a part is played by guiding words that can only gradually be understood, you might be reminded of an image we can use when we want to characterize the part of the fifth post-Atlantean cultural era that is now drawing to a close. In many respects it can really be compared with the season of Advent where the periods of daylight grow shorter and shorter. So now, in our time, when we can begin to have knowledge of revelations of the spiritual worlds again, evolution is entering the phase that we can picture as the days growing longer and longer, and we can speak of this season really being comparable to the 13 days and to the time of increasing daylight.

But it goes deeper than this. It would be absolutely wrong if we were only to find bad things to say about the materialistic age of the past four centuries. Modern times were ushered in by what the materialistic age has called the 'great' discoveries and inventions, sailing round the world, for instance, discovering lands that were not previously known and starting to colonize the earth.

That was the beginning of materialistic civilization. Then the time gradually came when people were almost stifled by materialistic civilization. The time arrived when all our spiritual forces were applied to understanding and grasping material life. Insights, understanding and visions of the spiritual world existing in ancient knowledge were forgotten more and more, as we have seen.

Yet it is wrong to have nothing but bad things to say about this age. It would be far better to put it this way: 'The human soul has been thinking materialistically and founding a materialistic science and culture in the part of it that is awake, but this human soul is a totality.' If I wanted to put it in a diagram I could say that one part of the human soul founded materialistic civilization. Before that, this part was inactive, and people knew nothing about external science and outer material life; at that time the spiritual part was more awake [drawing]. During the past four centuries the part of the soul was awake that founded materialistic civilization, and the other part was asleep. Yet, in truth, the age of materialistic culture was the time when the seeds were being sown in the sleeping parts of the soul for the forces we can now develop in humanity to bring us to spirituality again. During these centuries mankind was really an Olaf Åsteson as far as spiritual knowledge was concerned. That really was so, and humanity has not yet woken up. Spiritual science must awaken it. A time must come when both young and old must hear the words that are being spoken by the part of the human soul that was asleep in the age of darkness.

The human soul has slept long, indeed, but world spirits will approach and call out to it, 'Awaken now, O Olaf Åsteson!' Only we have to prepare ourselves in the right way, so that it doesn't happen that we are faced with the call, 'Awaken now, O Olaf Åsteson!' and have not the ears to hear it. That is why we are engaged in spiritual science, so that we shall have the ears to hear, when the call to be spiritually awake sounds in human evolution.

It is a good thing if we remember sometimes that we are a microcosm and that we can be receptive to certain experiences if we open ourselves to the macrocosm. As we have seen, the present season is a good one for this. Let us try to make this New Year's Eve a symbol for the New Year's Eve that has to come to mankind in Earth evolution, a New Year's Eve that will herald a new era bringing ever more light, soul light, vision, knowledge of what lives in the spirit and which can stream and flow into human souls from out of the spirit. If we can bring the microcosm of our experience on this New Year's Eve into connection with the macrocosm of human experience over the whole earth, we shall then have the kind of feelings we ought to experience, sensing as we do the dawning of the great new Cosmic Day in the fifth post-Atlantean era, a day at whose beginning we stand as we experience with dignity the midnight that precedes it.

LECTURE 5

Moral Experience of Colour and Music

Dornach, 1 January 1915

Everything in the world, each event and every act of human conduct, has two sides, which are like polar opposites.

Yesterday it was my task to point out to you that by understanding our spiritual-scientific world outlook through their feelings, human souls should acquire reverence and devotion for the spiritual worlds. The opposite pole to this reverent attitude is energetic work on our inner life, an energetic taking-in-hand of the evolutionary factors of our own soul, making a point of always using our experiences for the purpose of learning something from them and making progress with regard to our inner powers, so that whatever meets us in life and whatever takes place around us, whether it is easy to understand or not, we shall always avoid the danger of losing ourselves. May we always have the chance of keeping hold of ourselves and of finding within ourselves the strength to develop an understanding for what can come to meet us in an often incomprehensible way, and may the kind of reverence for things we were talking about yesterday make such an impact on the development of our souls that we acquire a proper understanding of universal existence; this, my dear friends, is the New Year greeting I wanted to give you today at the start of the year.

Having remembered reverence yesterday, I would like to follow that with a reminder of the energetic work on our inner life. The full moon shining down on us out of the cosmos on New Year's Eve yesterday was a symbol for the sequence in which we are remembering these things. If it had been the other way round, and the year had started with the new moon, I should have done the right thing in bringing these reminders to you in the opposite order. I should then have closed the year yesterday with a reminder of the power of inner development and should have had to leave the thoughts on reverence until today.

We must attach more and more importance to really noticing a symbol like this that shines on us out of the macrocosm. In our quiet moments this year let this indication work on us in such a way that it can have a special significance to think first about what transformation the power of reverence can bring about in us, and then to follow this with thinking about the transformation that the power of inner preservation, the maintaining of inner soul energy, should bring about in us.

The sequence of these thoughts has been set for the coming year by the writing of the stars, and the world will gradually come to realize again that reading the star script really has a meaning for mankind. So let us try even in details like this to pay heed to the great law of human existence and to strive for harmony between the macrocosm and the microcosm. The macrocosm is speaking to us over these days in the most obvious way, and we shall find how to keep our microcosm in harmony with the macrocosm if we conduct ourselves in line with it, in the course of this year that has begun amid such painful events.

If you try and notice what has been the prevailing mood of our talks in the last few days you will discover that, regarding the facts that spiritual science has made significant for us, we live in a time of change, even a time of hope, for we should now be getting an inkling of the direction the further course of human cultural development should take, the change that has to come about from a purely materialistic world conception to a spiritual one. What is meant here, however, cannot fully take place unless it enters every sphere of life and particularly takes hold of the spiritual, cultural areas to a marked degree.

We have already had various indications showing that an understanding of spiritual science, which has taken hold of our feelings and is not merely rational, is bound to bring a change both in artistic creativity and artistic appreciation, because the forces we gain from this can give us an artistic conception of the world. The very attempt we have been making with our Goetheanum building has been to demonstrate at least a small part of what can take on an artistic form through impulses from spiritual science.

We can see a time coming when we shall be able to enter fully into the feelings that can arise for us from the world view of spiritual science, a time when the way to artistic creation will in many respects be different from that of the past; it will be much more alive and the medium of artistic creation will be experienced much more intensely in the human soul; the soul will be capable of experiencing colour and music far more inwardly, in a kind of moral-spiritual way, and in artists' creations we shall to some degree meet traces of those artists' experiences in the cosmos.

In essentials, the attitude of artistic creation and artistic appreciation in times gone by was a kind of external observation, an appeal to something that affects the artist from outside. The need to refer to nature and to a model for external observation has become greater and greater. Not that in the art of the future there is to be any one-sided rejection of nature and outer reality. Far from it, for there will be a much more intimate union with the external world; so strong a union that it will cover not merely the external impression of colour and music and form, but also the impression one can experience behind the music and colour and form, what it is that colour, music and form reveal.

Human beings will make important discoveries in the future in this respect. They will actually unite their moral-spiritual nature with the results of sense perception. An infinite deepening of the human soul can be foreseen in this domain. Let us pick a particular example to start from. We will simply imagine that we are looking at a surface shining all over with the same shade of brilliant vermilion, and let us assume we succeed in forgetting everything else around us and concentrate entirely on experiencing this colour, so that we have the colour in front of us not merely as something that works upon us but as something with which we ourselves are united, something inside which we reside. You will then be able to feel as though you were in the world, and in this world the whole of you has become colour, the whole of your innermost soul, and wherever your soul goes in the world you will be a soul filled with red, living in, with and out of red. Yet you will not be able to experience this intensely in your soul unless the corresponding feeling is transformed into moral experience, real moral experience.

If we were to float through the world as though we were red, if we had identified ourselves with red, and our very soul and the whole world were entirely red, we should not be able to help feeling that this whole red world were filling us with the substance of divine wrath, coming towards us from all sides in response to all the possibilities of evil and sin in us. In this infinite red space we should be able to feel as though we were before the judgment of God, and our moral feeling would become a kind of moral experience our souls can have in infinite space. When the reaction then comes, when something emerges in our soul as we are having this experience in infinite red or, I could also say, in the one and only red, I can only describe it by saying we learn to pray. If you can experience in the red the raying and glowing of divine wrath, together with all the possibilities of evil in the human soul, and if you can experience in the red how one learns to pray, the experience of red is enormously deepened. Then we can also experience how red can create form when it enters space.

We can then understand how we can experience a being that radiates goodness and is full of divine kindness and mercy, a being that we want to bring through feeling into the realm of space. Then we shall feel the need to let this divine mercy and goodness express itself in a form that arises out of the colour itself. We shall feel the need to let space be pushed aside so that goodness and mercy may shine forth. Before space was there it was all concentrated at the centre, and now goodness and mercy enter the space and, just as clouds are driven apart, space is rent asunder and recedes to make way for mercy, and we have the feeling that what is being scattered must be drawn in red. Here in the centre [drawing] we shall have to indicate faintly a kind of magenta shining into the scattering red.

We shall then be present with our whole soul as the colour takes on form. With our whole soul we shall feel an echo of how the beings who belong especially to the coming into existence of our earth felt, when they had ascended to the Elohim stage and learnt to fashion the world of forms out of colours. We shall learn to experience something of the creative activity of the Spirits of Form who are the Elohim, and we shall then understand how colour can create forms, as indicated in the first Mystery Drama.[26] We shall also begin to understand how the surface nature of the colour becomes something that has to be overcome, as it were, because we accompany colour into the cosmos. If strong desire is also present, a feeling can arise like Strader has when he sees the portrait of Capesius, and says he would like to pierce through the canvas.

You will see that an attempt has been made in these Mystery Dramas really to show in artistic form what it is like for the soul when it attempts to expand in the cosmic forces and to feel what the spirits of the cosmos are feeling. That was indeed the beginning of all art. But the materialistic age had to come, and ancient art, with its divine nuances, in which spirit as something inward was revealed in matter, had to change into secondary, materialistic sham-art, which the art of the materialistic age is, in essence; the kind of art which cannot create but only copies. The sign of all second-hand art, all sham-art, is that it needs objects to imitate; it does not produce first-hand forms by working directly with the material.

Let us take another example. Let us imagine we do the same thing with a more orange coloured surface that we did with the red surface. We shall experience something quite different this time. If we immerse ourselves in it and unite with it then, instead of feeling divine wrath bearing down upon us, we shall have the feeling that what comes to meet us here still has perhaps a little of the commanding aspect of wrath about it but that it does not want to punish us so much as to impart itself to us and arm us with inner strength.

When we enter the world and become one with the orange surface we move in such a way that with every step we take we feel that by experiencing orange, by living in the forces of orange, we are becoming stronger and stronger, and that what comes to us out of orange does not come merely to punish us and break us with its judgment, but is a source of strength. This is how we go into the world in orange. We then feel the longing to understand the inner nature of things and to unite it with ourselves. By living in red we learn to pray, by living in orange we experience the desire, the longing, for knowledge of the inner essence of things.

If we do the same thing with a yellow surface we feel as though we were transported back to the beginning of our cycle of time. We feel that we are living in the forces out of which we were created when we entered upon our first earthly incarnation. You feel an affinity between what you are through the whole of earth existence and what comes to meet you from the world into which you take the yellow with which you are united.

If you identify with green and accompany green into

the world—which can be done very easily by letting your eyes roam over a green meadow, shutting out everything else and concentrating completely on it, and then trying to immerse yourself in the green meadow as if the green were the surface of a coloured sea—you experience an inner increase in strength in what you are in a particular incarnation. You feel yourself becoming inwardly healthy, yet at the same time becoming inwardly more egoistic; you feel a stimulus of the egoistic forces within you.

If you were to do the same with a blue surface, you would go through the world with the desire to accompany the blue for ever and to overcome your egoism, become macro-cosmic, as it were, and develop devotion. You would find it a blessing if you could remain like this for your meeting with divine mercy as it comes towards you. You would feel blessed by divine mercy if you could go through the world like this.

In this way we learn to recognize the inner nature of colour and, as I said before, we can foresee a time when an artist's preparation will mean a moral experience in colour of this kind, when the experience preparatory to artistic creation will be much more inward and intuitive than it ever was in past ages. I am only giving you a few indications, and these will be developed much further in the future. They will take hold of people's souls and enliven them with a tremendous sense for artistic creativity, whereas the materialistic culture that has entered our modern age has dried up the soul and made it passive. Souls must be stimulated again by a power from within; they must be taken hold of by the inner forces of things. I have taken the colours that flood the world as a specific example.

The world of music will deepen and enliven the life of soul in a very similar way. During the period that is now drawing to a close the essential thing was that a person experienced a musical note as such, and the relationship of one note to another. In the future people will be able to experience what is behind the note. They will regard the note as a kind of window through which they enter the spiritual world, and then it will not depend on vague feelings of how one note is added to another, forming melodies for instance, but by going through the note the soul will also experience a moral-spiritual quality behind the separate notes. The soul will enter the spiritual world as though through a window. The secrets of the individual musical notes will be discovered behind the notes when they are experienced like this.

We are still a long way from this feeling of being able to go from the sense world to the spiritual world through the window of each note. But this will come. We shall experience the note as an opening made by the gods from the spiritual world to this physical-material world, and we shall climb through the note out of the physical-material world into the spiritual world.

Through the prime, for example, which we experience as absolute and not in reference to previous notes of the scale, we experience danger as we pass from the sense world into the spiritual world. We are threatened on entry with being taken captive; the prime wants to suck us in most horribly through the window of the note and make us completely disappear in the spiritual world. Assuming that we experience the prime as absolute, we shall feel that we are still too weak in a spiritual sense in the physical world, and that we are sucked up by the spiritual world when we climb through this window. This is the moral experience to be undergone on entering the spiritual world through the prime. I am over-simplifying it now, though; in actual fact we shall have a very differentiated experience which contains an infinite variety of detail.

When we climb out of the physical world into the spiritual world through the window of the second, we shall gain the impression of powers in the spiritual world that take pity, as it were, on our weakness and say, 'Well! so you were weak in the physical sense world! If you only climb into the spiritual world through the prime I must dissolve you, suck you up and break you to pieces. But if you enter through the second I will offer you something from the spiritual world and remind you of something that is there as well.' The peculiar thing is that when we climb from the physical to the spiritual world through the second, it is as though a number of notes were ringing out to receive us. We enter a totally silent world if we enter the spiritual world through the absolute prime. If we enter through the second we come to a world where, if we listen, notes of varying pitch resonate gently to comfort us in our weakness. Thus we have to go through the window into the spiritual world in a way we most certainly could not do in a physical house, for the owner would give us a strange look if we were to walk in through the window and bring the whole window with us. But in the spiritual world that is what we have to do, we have to bring the window with us, bring the notes with us and, in union with them, live over there in the beyond on the other side of the thin partition that then separates us from the physical-material world.

If you enter the spiritual world through the third, you will have the feeling of an even greater weakness. If you enter the spiritual world this way, you will feel that you are really very weak in the physical-material world, where its spiritual content is concerned. But with regard to the third—and remember that you have become the notes, you yourself have become a third—you will feel that there are friends over there who, although they themselves are not thirds, approach you according to the kind of disposition you had in the physical-material world. If you enter through the second, it is like a gentle resonating of many notes, with whom you share life in general when you enter through them, whereas notes that are like friends with one another come to meet you through the third. People who want to become composers will have to enter especially through the third, for that is where the tonal sequences, the tonal compositions are that will stimulate their artistic creativity. You will not always be met by the same tonal friends, for which ones you meet will depend on your mood, feeling and temperament—in fact how you are disposed to life at the time when you go through the third into spiritual life. This results in an infinite variety of possibilities.

If you penetrate into the spiritual world through the fourth you will experience something strange. Although no new notes appear from any direction, those that have come before, when you were experiencing the third, will live lightly in the soul as memories. You will find that in continuing to live with these note-memories they perpetually take on a fresh colouring; at one time they become as bright and cheerful as can be, at other times they descend to the utmost sadness; now they are as bright as day, now they sink down to the silence of the grave. The intoning of the voice, the way the sound ascends and descends; in short, the whole mood of tonal creation will have its origin along this path, from these musical note-memories.

The fifth will produce experiences that are more subjective, that work to stimulate and enrich the life of the soul. It is like a magic wand that conjures up the secrets of the music world over there, out of unfathomable depths.

Experiences of this kind will come to you when you no longer merely look at the things and phenomena in the world, or merely listen to them, but experience them from within. These are the kind of experience you have to have, particularly through colours and musical sounds, but also through form, in short, altogether through the realm of art, in order to get away from the purely external relationship to things and their functioning—which is characteristic of the materialistic age—and to penetrate into the inner secrets at the heart of things.

Then something tremendously significant will happen to human beings, and they will be filled with the awareness of their connection with the divine spiritual powers, which are subconscious when their awareness is confined to the material realm, and which guide and lead them through the world. Then, above all, they will have inner experiences, such as experiencing the forces which guide us from one incarnation to the next. If we omit to stoke up a locomotive, it cannot pull a train. The forces that make things happen in the world have to be continually stimulated. The forces that drive mankind forward also need to be stimulated, and this does happen. However, we have to learn that we are connected with these forces.

I once had the following remarkable experience. There was a lawyer, a famous advocate, in the town where I lived for a while, an extraordinarily famous lawyer to whom people flocked because they believed he was bound to win the most difficult lawsuits, which was indeed often the case.[27] His legal dialectic was extraordinary, and people who knew him had the deepest respect for it. On one occasion he was entrusted with the difficult lawsuit of a rich man. The rich man would have to pay a heavy fine if the case resulted in his being sentenced. The lawyer used his greatest eloquence and the most wonderful of his legal skills. He made a long speech, and the people who heard him were convinced that if this did not persuade the jury to acquit the accused, nothing else could have done so either. Everyone who heard the lawyer's amazing skill was absolutely convinced the jury would now withdraw and acquit the accused.

But there was in the law court not only a skilful counsel but also a skilful judge, and although the hour was not yet so advanced as to leave too little time for the judgment, the judge said, 'Let us adjourn for today and continue tomorrow.' So the jury's session was to take place the following morning, and this gave them time to think the matter over again during the night. The following day came. This 'overnight delay', as he called it, had made the lawyer feel very uncomfortable. The session began, the jury withdrew, and everyone awaited their return with tense anticipation, most of all our lawyer. The jury came back after only a quarter of an hour, and when the lawyer heard them coming back from the conference room so soon, he fainted. Yes, he fainted. He did recover again, and a friend helped him up. And the accused had been convicted, but the lawyer only heard this after he had recovered from fainting.

What might be said about such a course of events

looked at from the point of view of external perception? One might have said that the lawyer was a very ambitious man, for he cared so much about winning this lawsuit that he lost consciousness before the verdict was pronounced. As soon as he saw that the jury had only conferred for a short time he was sure the accused would be convicted, for if they had acquitted him they would, of course, have taken much longer.

But that was not how the matter stood; it only appeared to be so from an external point of view. There was a kind of second layer of events behind what I have just told you. This other layer, or story, is like this: Before the court case took place the lawyer had become addicted to what one can call the demon of gambling—as I said, I knew him well. He played the stock exchange using sums of money entrusted to him as deposits. It was an absolute passion with him, and not long before this lawsuit began he had reached the point where he had gambled away large sums of money entrusted to his care. But he had been promised that if he should win this suit he would be paid so much that he could more or less cover his losses.

So he did not faint only out of wounded ambition; he fainted when the jury returned with the verdict after only a quarter of an hour, because his livelihood had in fact been destroyed, for he had no hope any more of replacing the deposit money he had lost. His whole livelihood had depended on the outcome of the lawsuit. He fainted as symbolic indication that he was now completely ruined for this incarnation. He had to escape to America after that, and had to endure a not very enviable existence there for the rest of his life.

An example of this sort shows how wrong our judgement can very often be, for some people will never have heard anything about what had been going on behind the lawsuit. If these people had only heard the clever lawyer during the court case and seen him fainting, they could very well have concluded that some people are so ambitious that they lose consciousness if a speech they give misfires. And they would have left it at that.

To be able to judge correctly in this instance you had to be aware of a further layer of facts. In many instances you would have to know several more layers, otherwise you could be correct with regard to the layer you can see and yet make a wrong assessment. That is the external aspect. But the matter has still more behind it. This fellow also had to find a way of getting from this incarnation to the next one, and here we have an example of how wise world guidance puts into the soul the forces it needs to lead it from one incarnation to the next. The man was in such existential conflict that it had destroyed his very existence. He was in a terrible position. A situation had been created in which there were no forces left to carry him over to his next incarnation. Also, a situation had been created in which forces of that nature could not be brought into his conscious mind. So his consciousness had to be extinguished for a moment. During short breaks in consciousness all kinds of other spirituality can enter the human soul. In that moment he received forces capable of restoring his impulse to go forward into the next incarnation. Of course an impulse like this can be given in many different ways. What I have described here was one particular case.

These impulses are always there. But I just wanted to show you that our conscious life is linked with an ongoing process in the unconscious, and that in conscious life there are actually points where the consciousness is suppressed so that something can enter out of the unconscious. Sometimes these unconscious moments need not be long; they can be short spells similar to fainting. Yet a tremendous amount of spiritual vitality can stream into the human being at such moments, forces that may be good or bad, and capable of either.

What I want to show you with this example is that, in observing the world, we must try to notice links of this kind, which are of no significance to a materialistic outlook.

You will gradually reach the point of becoming so perceptive for living links that you will recognize the moments in which the spirit comes near to each human being. In the future the world will no longer be described as something so undifferentiated as it is nowadays on the basis of material causes; matter will be relegated to its proper place and, at the same time, people will realize that the material phenomenon is not the only thing, for spirit shines through it.

We have seen that colours and musical notes are windows through which we can ascend spiritually into the spiritual world, but life also brings us windows through which the spiritual enters our physical world. The lawyer's fainting fit was this kind of window. If we interpret this event correctly we have to say that spiritual life streams down to us through this window. It is clear to us that these forces flowing into us cannot be explained on a purely material level. So there are windows in the notes through which we ascend from the physical-material world into the spiritual world; and there are also windows through which, if we remain in the physical material world, the spirit can descend to us.

If we fail to perceive the fact that spirit descends to us through such windows, it is like someone who cannot read opening a beautiful book. He has the same thing in front of him as someone who can read, but if he cannot read he sees unintelligible scribbles on the white pages which he cannot interpret although perhaps he can describe them. Only a person who can read is capable of following a biography, perhaps, or a piece of information that has been set down in these strange signs. A person who cannot read world phenomena is like a cosmic illiterate where these phenomena are concerned. A person who can read, however, reads the ongoing process of the spiritual world in them. It is characteristic of the present materialistic age that materialism has made people illiterate with regard to the cosmos, almost a hundred per cent so. At a time when people are so proud of having reduced the percentage of illiteracy in civilized countries to such a great extent, they are enthusiastically heading towards illiteracy where the cosmos is concerned.

It is the task of spiritual science to eliminate this cosmic illiteracy. Few people nowadays are illiterate in the ordinary sense. In the time of ancient clairvoyance human beings were far less illiterate in the spirit. But this must not make us conceited. It is a fact that when we acquire an inkling of our task in the spiritual-scientific stream, we ought to change from being illiterate to becoming people who can read the cosmos. Yet we should remain humble, for the times are such that we are still very much in need of the elementary level of education. We can hardly read yet, but only spell out the letters. Yet we can be gripped by the impulse to change, an impulse which is breaking in upon mankind through these things, and if we are gripped by them we shall have the right attitude to what the signs of the times are demanding of us, and shall enter into them as people who rightly belong to the stream following the world view of spiritual science.

LECTURE 6

Working with Sculptural Architecture—I

Dornach, 2 January 1915

It will be relatively easy—and I mean relatively—for a person to take up more or less theoretically what we understand by the world outlook of spiritual science or anthroposophy. But it will not be easy to fill our whole being and life itself with the impulses coming from spiritual science. To absorb anthroposophy theoretically—so that you know the human being consists of a physical body, an etheric body, an astral body and so on, in the same way as you know that one note or another has vibrations at such and such a frequency, or that oxygen combines with hydrogen to make water—is the way we have grown accustomed to learning things through the scientific approach that has gradually developed over the last few centuries.

We are less accustomed, however, to allowing our feelings and attitude of mind to be affected by the kind of knowledge spiritual science has to offer. Yet the kind of approach we must have to spiritual science is fundamentally opposite to the approach needed for ordinary science. It is often stressed that everyone feels science to be dry and that it stops us having a warm contact with life and its happenings; that dry science has something cold and unfeeling about it and robs things of their dewy freshness. Yet one could say that to a certain extent this has to be so with science. For there is an enormous difference between the impression made on us by a wonderful cloud formation in the evening or morning sky and the bare reports an astronomer or a meteorologist gives us about it. There is such abundant richness in the natural world around us that its effect on us is to warm us through and through whereas, in comparison, science, with its concepts and ideas, appears dull and dry, cold, lifeless and loveless. There is every justification to feel this way as far as external scientific knowledge is concerned.

There are good reasons why external, scientific knowledge has to be like this, but spiritual science is not this kind of knowledge. On the contrary it ought to bring us nearer and nearer to the living abundance and warmth of the outside world and of the universe altogether. But this means learning to bring certain impulses to life in us that a person of today hardly possesses at all. A present-day person expects it to be in the nature of what he calls science that it has a cold and sobering effect on him, like the character of Wagner in Goethe's *Faust*. He expects that if he assimilates science the riddles of nature will be solved, and then he will know how everything is constituted and be perfectly satisfied with what he knows.

Science even makes some people shudder nowadays, for a quite specific reason. They maintain that what made life so rich and fresh in the past was the fact that man had not solved every riddle and could still wonder about the unsolved ones. Now science comes along, they say, and solves the riddles of nature one after the other. They imagine how boring life will be in the future when science will have solved all the mysteries and there will be no further possibility of wondering about anything, or having any feelings of an unscientific kind. What terrible desolation would befall mankind; we have every reason to be horrified at the prospect.

Spiritual science, on the other hand, can kindle different feelings, and although such feelings would be less in keeping with modern times than the solving of riddles, they show how awakening and life-giving spiritual science can be. If we absorb spiritual science in the right way—if we do not merely take notes of what is being said so as to make use of them as people do in ordinary science, perhaps even doing a neat diagram so as to take it all in at a glance like physics; if we do not do it so much that way, but let what spiritual science has to say reach our hearts; if we unite with it, we shall notice that it comes to life in us and grows, it awakens our independence and initiative and becomes like a new living being within us, that is for ever showing new aspects. To approach external nature with our souls thus filled with anthroposophy is to find more riddles in nature and not fewer. Everything grows even more puzzling, which broadens instead of impoverishing our life of feeling. You could say that spiritual science makes the world even more mysterious.

The world cannot help becoming a desolate place when the physicist says to us, 'You see the sunrise...' and then, showing us a diagram, proceeds to tell us which particular refractions are taking place in the rays of light so that the glow of dawn appears. This is certainly horrible, not from the point of view of human reason, but for the human heart and an understanding connected with the heart.

It is quite different when spiritual science tells us, for example, 'When you see the sunrise or hear a piece of music, it must feel to you as though the Elohim were sending their punishing wrath into the world.' Then we become aware of the mysterious living weaving of the Elohim behind the glow of dawn. Yet to know the name of the Elohim and to be able to give them a place in the ninefold diagram we have drawn in our notebook, is not to know anything about the Elohim. But out of the living feeling we can have in looking at the sunrise will come a perception of movement and life in rich abundance, just as we know that when we look at a human being, any amount of conceptual knowledge about him will not tell us the whole of his nature, nor fathom the universal life within him. Likewise, we become aware that the dawn is revealing something to us of the unfathomable life of the cosmos.

The knowledge of spiritual science makes life more enigmatic and mysterious in a way that kindles richer feelings within us. It is a fundamental feeling of this kind our souls can acquire when we bring spiritual science to life within us and when we try to make ourselves at home in the kind of ideas I have just indicated. Then we shall never be tempted to complain that spiritual science only appeals to our heads and does not take hold of our whole being. We just need to be patient until the message of spiritual science becomes a living being within us and forms itself anew, filling us not only with its light but also with its warmth. Then it will take hold of our hearts and our whole being and we shall feel the richer for it, whereas if we take up spiritual science in the same way as ordinary science we are bound to feel the poorer.

On the other hand, it is quite natural that initially spiritual science seems to many people to lead to an impoverishment, because they have not yet been able to find the inner life of the message of spiritual science that can reach their heart, and because spiritual science does not yet have the same effect on them as, for instance, the warm words of a fellow human being speaking to them. We have to learn to let spiritual science come alive, so that it can give us as much support and encouragement as we can otherwise only receive from another human being.

Our hearts find this so difficult at the present time because we have lost the habit of uniting ourselves with the life of things. It is difficult enough if one tries even in small doses to re-introduce this living with things. This was attempted in the four Mystery Dramas. You only have to think of the scene in spirit land, the fifth scene of The Souls' Awakening,[28] where Felix Balde is sitting on the left side of the stage—as seen from the audience— having ascended to Devachan, and a spiritual being on the other side of the stage speaks to him of its experience of weight. Here one should feel the heaviness that is floating down in the distance. When people see something floating down they are accustomed nowadays only to be aware of the descending and only to see the thing higher up to start with, and then coming lower and lower down. They are quite unaccustomed to getting inside things and feeling the experience of weight, feeling the heaviness pressing down on things all the time. With a passage like that in the middle of the play I was hoping to lift people out of their egoistic bodies and put them into the life of things outside themselves.

If this cannot happen, then real artistic feeling will not be able to arise again. For instance, in order that a true feeling for architecture can come again, the concepts we receive from spiritual science must come alive. To begin with it makes very little difference which particular anthroposophical concepts we carry around with us. But if we really do something like this we shall see how much richer our soul life becomes. For example, we shall gain a lot if, as well as just seeing this diagram, we try and get inside it and try to feel what is going on: weight bearing down here, and weight being supported there.

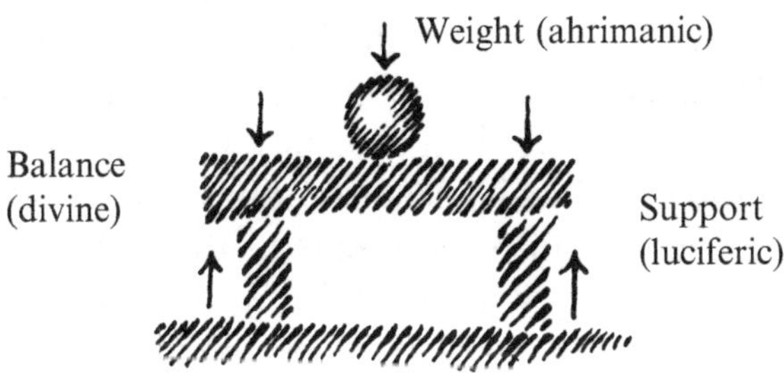

We must go further than merely looking at it; we must feel that the beam needs to have a certain strength, otherwise the load will crush it, and the supporting columns must also have a certain strength, otherwise they too will be crushed. We must feel the way the sphere on top is weighing down, the columns are supporting and the way the beam is keeping the balance. Not until we get inside the elements of weight and support, and inside the balance between the weighing down and the supporting, shall we feel our way into the element of architecture.

But if we follow a structure of this kind not only with our eyes but, as it were, get inside it and experience the weighing down, the supporting and the balance, then we shall feel that our whole organism is becoming involved, and as if we have to call on an invisible brain belonging to our whole being and not just our head. Then we can become aware, 'Ah! now we are beginning to feel!' To take our simple example, we shall feel a supporting element, an upward striving, supporting luciferic element; a weighing and bearing down ahrimanic element; and between the luciferic and ahrimanic elements a balance which is a divine quality. Thus, even lifeless nature becomes filled with Lucifer and Ahriman and their superior ruler, who eternally brings about the balance between them.

By learning to experience the luciferic, ahrimanic and divine elements in architecture, so that architecture affects us inwardly, we shall become conscious of a richer feeling of the world which leads or, one could almost say, draws the soul out into the things of the world; for our soul is now not only within our body's skin but belongs to the cosmos. This is a way of becoming conscious of this. We shall become aware, too, that whereas outside the architectural element is supporting, weighing down and creating a balance, in this encounter with the architectural element we ourselves develop a musical mood. Our inner being becomes attuned to architecture in a musical mood, and we notice that even though the elements of architecture and music appear to be so alien to one another in the outer world, through this musical mood engendered in us, our experience of architecture brings about a reconciliation, a balance between these two elements.

This is where, from our era onwards, living progress in the arts will lie, through learning to experience the reconciliation of the arts. This was dimly felt by Wagner, but it can only really come about when the world comes alive with spiritual science.

Reconciling the arts: that is what we have attempted to do—for the first time, and in a small, elementary way—

in our Goetheanum building in which more is to happen than cold and abstract chatter about this reconciliation, for the building itself is intended to show in its architecture an impression, a copy of this reconciling of a musical mood with architectural form. If you study what is presented in the sequence of columns and everything connected with them, you will discover that we were making the attempt to bring the elements of support, weighing down and balance into living movement. Our columns are not merely supports, and our capitals no longer mere supporting devices, and the architraves that extend above the columns do not just have the character of rest, serving only to round the columns off at the top, but they have a character of living growth and movement.

We attempted to bring architectural forms into musical flux, and the feeling one can have from seeing the interplay between the columns and all that is connected with them can of itself arouse a musical mood in the soul. It will be possible to feel invisible music to be the soul of the columns and the architectural and sculptural forms that belong to them. It is as though a soul element were in them. Interpenetration of the fine arts and their forms by musical moods has to be the fundamental ideal of the art of the future. Music of the future will be more sculptural than music of the past. Architecture and sculpture of the future will be more musical than they were in the past. That will be the essential thing. Yet this will not stop music from being an independent art—on the contrary, it will become richer and richer through penetrating the secrets of the notes, as we said yesterday, creating forms out of the spiritual foundations of the cosmos.

However, since everything that is inside must also be

outside in art, since all that lives in it must be embodied as though in an organism, the world of soul within the series of columns and everything belonging to them must also become embodied. This happens, or rather is intended to happen, in the painting of the domes. Just as the columns and everything belonging to them are like the body of our building, so, when you are inside the building, is all that is going to appear in the domes its soul; and just as the world appears to be filled with spirit, when our organs are directed outwards, so shall our windows executed in the new art of glass shading represent the spirit. Body, soul and spirit shall be expressed in our building. Body in the column structures, soul in everything to do with the domes, and spirit in what is created in the windows.

Where all this is concerned karma has brought about various things for which we can be grateful, for precisely in the case of the Goetheanum building, karma has indeed helped us in several matters. The soul of a human being is so constituted that from outside we perceive it in his physiognomy, but we have to have resources like love and friendship to penetrate into a person's soul if we want to get to know it from the inside.

When I was travelling from Kristiania to Bergen on my last lecture tour in Norway, I happened to see a slate quarry which gave me the idea of trying to get slate from there. We were successful, and it really was what one might call a stroke of karma, for when we look at the roofs of the domes that are now covered with this slate with its unique qualities, we have to admit that they have something of the quality of the life of the soul, which at one and the same time both discloses and conceals what is within. Moreover, if we really want to feel the domes as soul life we shall have to develop a love for spiritual science, for what is going to be painted inside the domes should really appear to us as a kind of reflected image, in colour and form, of what spiritual science can mean to us. To see this we have to go inside. But when the building is really finished, if it ever is, no one will be able to understand what he sees when he goes inside if he has not developed a love for spiritual science; otherwise what he sees there will probably remain something that can perhaps cause some sensation, but will not be anything that particularly appeals to his heart; what he gets from it will easily tempt him to deny that the architecture has anything to offer the feelings.

Just as we have seen in this instance that what comes to life out of spiritual science can be rediscovered in the world, so can life be made fruitful through spiritual science in realms where we can more readily see that our heart's understanding needs to be warmed and made alive. For it is not only artistic and scientific areas that are to be made fruitful by spiritual science, but the whole of life has to be.

As an example let me take a realm in which we can see particularly well how anthroposophical concepts can come alive in outer life. I will choose the realm of education, any kind of art of education. Let us begin from the fact that children are educated by grown-ups. What does the materialistic age envisage when it speaks of a child being educated by an adult? Fundamentally speaking, the materialistic age sees in both of them, both the adult and the child, only what you get from a materialistic outlook, namely, an adult teaching a child. But it is not like that. Externally the grown-up is only maya, and seen from outside the child is only maya too.

But there is something inside the grown-up that is not directly contained in the maya, namely, the invisible human being who passes from one incarnation to another, and there is also an invisible part in the child that goes from incarnation to incarnation.

We shall speak about these things again. But I would like to tell you a few things today from which you will see in the course of time—if you meditate on them—what else there is in spiritual science. I will start with the fact that a person, as he appears in the external world, cannot teach at all, nor can the person who stands before us, externally, as a child, be educated. In reality something invisible in the teacher educates something invisible in the pupil. We shall only understand this properly if we focus our attention on what is gradually unfolding in the growing child, as the outcome of previous incarnations. When everything coming from previous incarnations has made its appearance, no more education is possible, the child withdraws, especially in present times. What we are actually educating is the invisible result of previous incarnations. We cannot educate or have any influence on the visible child. That is how the matter stands with regard to the child.

Now we will look at the teacher. During the first seven years of the child's life the teacher can only educate by means of what the child can imitate; in the second seven years it will be through the influence he has as an authority; and finally in the following seven years it will be through the educational effect of independent judgement. Everything that is active in the teacher all this time is not in his external physical part at all. The part of us which does the educating will not take on physical form until our next incarnation. For all the qualities in us which can be imitated, or the qualities upon which our authority is based, are germinal qualities and will form our next incarnation. When we are teachers our own next incarnation converses with the previous incarnation of the pupil. It is an illusion to think that as present people we speak to the child now present before us. We only have the right feeling for this if we say to ourselves, 'The very best in me which my spirit can think and my soul can feel, and which is preparing itself to make something of me in the next incarnation, can work on the part of the child that is sculpturing its form out of times long past.' What enables us to educate is something intrinsically musical in us. What we should work upon in the child is something that is doing sculpture in that child.

Take as a whole all that I have said in these lectures about the musical element and how, in its most exalted form, it corresponds to what the human being meets with in initiation. Music is related to everything that is in a process of development and lies in the future, and the realm of sculpture and architecture is related to what lies in the past. A child is the most wonderful example of sculpture we can see. What we need as teachers is a musical mood, which we can have in us in the form of a mood pervaded by the future. If you can have this feeling when you are involved in teaching, it will add a very special note to the relationship of the teacher to the child, for it will make the teacher set himself the highest aims while having the greatest measure of understanding, however naughty the children might be. There really is an educational force in this mood.

Once the world comes to see that the right atmosphere for teaching arises when a musical mood in the teacher is combined with seeing the sculptural activity in the pupil;

once it is established that this is what is required for a love of teaching, then education will be filled with the right impetus. For then the teacher will speak, think and feel in such a way that in the course of his lessons, what comes from the past will learn to love what reaches out to the future. The result will be a wonderful karmic adjustment between the teacher and his pupils—a wonderful karmic balancing.

If the teacher is egoistic and only tries to make the child an imitation of himself, then the teaching is purely luciferic. Education becomes luciferic when we try as far as possible to turn the pupil into a poor imitation of our own opinions and feelings, and are only happy if we tell the pupil something today and he repeats it word for word tomorrow. That is a purely luciferic kind of education. On the other hand an ahrimanic education comes about if the pupils are as naughty as possible in our lessons and learn from us as little as they possibly can. However, there is a state of balance between these two extremes, just as there is between weighing down and supporting. This is arrived at through the interplay of the musical and the sculptural elements I have just been speaking about. We must learn to distinguish between the teacher's intentions and what the child turns out to be. If we have the right mood, then even though we have been trying to teach our pupil something quite specific, we shall be overjoyed to realize that he has not turned out as we intended, but that the child has developed into something quite different from what we intended him to be. This is the remarkable thing, that the teacher can only rid himself of his egoism in teaching if he overcomes the desire to turn the child into a copy of his own views on what is good and right, and especially of his own favourite thoughts. The best thing we can achieve, as teachers, is to be able to face perfectly calmly the thought of the child becoming as different from us as possible.

But you cannot come along and say, 'Please give me a recipe for it, write a few rules down for me on how to teach like that.' The remarkable thing about the spiritual world outlook is that you cannot work according to rules, but you really have to absorb spiritual science, so that you are filled with it and your impulses of feeling and will are increased. Then the right thing will happen, whatever particular task you face in life. The essential thing is to tackle it in a living way.

You might ask, 'Which is the right teaching method from the point of view of spiritual science?' The correct answer would be, 'The best spiritual-scientific method of teaching is for as many teachers as possible to engross themselves in spiritual science in a living way, and to acquire the feelings that come from spiritual science.' This is less convenient, of course, than reading a text-book on the art of education based on spiritual science. Spiritual science is for ever being asked, 'What is the spiritual-scientific point of view on this or that?' Spiritual science, however, does not have a point of view, or, if you like, it has as many points of view as life itself. Spiritual science itself must become life. Spiritual science must be absorbed and brought to life within us, then it will be able to bear fruit in the various realms of life. People will then get beyond whatever it is that makes life so dry and dead: we could call it the desire for uniformity. External science requires uniformity, but spiritual science gives manifoldness and variety, the kind of variety that belongs to life itself. Thus, spiritual science will have to bring transformation into the furthest reaches of life.

Let us look at what some realms of life are like today. Learning takes place up to a particular age; you learn one thing up to one age and something else up to another. Then comes the time when you go out into life, as we say, and do not want to learn any more; even when you go in for a scientific career you do not like having to learn any more. The ones who do go on learning in order to keep up with their science are thought to be the odd ones. In the general run people learn until a certain age and after that they play cards or do other useless things in their spare time, or they develop an attitude like one I came across a while ago. I had been invited to give a series of lectures on the history of literature in a circle that included some ladies with a thirst for knowledge. It can be said that the more pliant or, if you prefer it, the 'less developed' brain that ladies have today is capable of retaining more of the receptivity and flexibility of ancient times, when learning continued throughout life. This is indeed more often found among women than men. The ladies in question had the feeling that they ought to bring the gentlemen along too. So the gentlemen were there, and they didn't all go to sleep. Some of them really listened. Afterwards there was conversation, tea and cakes, in other words they did what is considered to be obligatory in some circles if the lectures are not to be too dry. So there were these conversations, and after I had been lecturing on Goethe's *Faust*, some of the men summed up their attitude by saying, 'When you see *Faust* on the stage it isn't really the kind of art you can enjoy, it isn't even recreation, it is science.' This was their way of saying that when a person has been working in an office all day, or has been serving customers, or standing in a court of law interrogating witnesses and sentencing the accused, by the evening he is in no state to listen to Goethe's *Faust* any more and needs recreation and not science.

This is an example of a common attitude with which no doubt you are familiar. You only need to mention it, for everyone knows how widespread it is, and that a lot of people would find it strange the way we gather here in such a studious fashion and want to go on learning, despite the fact that several of us are fairly old. They think they know a much better way of spending time. Yet a complete change will have to take place in people's approach to spiritual science, in that they will not just want to let it remain a study, but must want to have a living and permanent relationship with it. This will come. You cannot learn spiritual science the same way you learn other sciences, by taking it down in a notebook; spiritual science must stay alive. It becomes dead if we only learn its content and do not remain connected with it through living activity. It becomes dead and withers away, whereas it should be kept alive. Spiritual science must work in this way to enliven us and keep our hearts receptive for all they can receive from the spiritual worlds, so that we develop further all the time.

There is no doubt that in our era humanity shows a quality of old age; on the whole it does not have the kind of youthfulness it had in mythical times. Spiritual science must be people's draught of youth, so that they will feel able to learn from life throughout their lives. Nowadays we can experience odd things in this connection. I know a man with an active mind, a person who has had all kinds of connections with modern intellectual culture all through his life.[29] He celebrated his fiftieth birthday recently, and published a pamphlet on this occasion containing some very peculiar notions. For instance he said—but I want to alter things a little bit, so that you won't guess who it is—he said, 'I have been offered a post in the realm of art that I had been longing for, for many years. But now that I have reached the age of fifty, old age in fact, I do not really want it any more. For to fill a post of that kind and to inspire the people around you, you need to be young, you need to be full of fantastic illusions. These illusions have to consist of thinking that what you are doing and the people you have to deal with are the whole world and nothing else matters. All that counts is what is right here. Fifteen years ago I was of an age when I could have done it. Now I am past it. You should not wait until people have grown old before you offer them influential positions, but let them be privy councillors, for instance, when they are between thirty and forty.' This was the gist of what this 'old' man said.

This mood is absolutely in line with the whole tone of our contemporary culture. It is a mood very easily acquired by people who accept what materialistic culture has to say about the human being, for materialism has not the power to penetrate the whole being of man; the content of this materialistic knowledge is not powerful enough to have the kind of influence on his soul life that will last right into old age. Spiritual science, however, wants to prove that even if people grow old externally they can stay young in soul, and if they have not done anything special by the age of fifty, although they do not need to succumb to the illusion that what they are doing is of prime importance and everything else can fall by the wayside, they can still be young enough to devote all their strength to what they have to do. They can be youthful, in fact childlike enough to concentrate the whole of their forces on what has to be done, just as a child concentrates all its forces in play. Spiritual science must become a magic draught of youth and not just a theory. That is also an impulse of transformation. Tomorrow I will talk about other impulses of transformation.

LECTURE 7

The Future Jupiter Evolution and its Beings

Dornach, 3 January 1915

When you recall the talks we had in connection with the evolution of Earth through the Saturn, Sun and Moon evolutions, you will remember that at each of these evolutionary stages one particular kind of being, from among what we would now call the higher hierarchies, attained its own human level. We know that during the Old Saturn evolution the Spirits of Personality, the Primal Beginnings, the Archai reached their human level, during the Old Sun evolution the Archangeloi, during the Old Moon evolution the Angeloi and during the Earth evolution, mankind.

You will also have seen from our talks on evolution that each level of beings that subsequently reached a certain stage of development first underwent preparation for this. We know that the human being was undergoing preparation throughout the Saturn, Sun and Moon evolutions, and that what we nowadays call man's completed physical body has been evolving since the Saturn evolution, the ether body since the Sun evolution, the astral body since the Moon evolution, and that the ego was only added to these during the Earth evolution. In this way the completed outcome of a particular group of beings is prepared.

You may be anxious to know whether, in our present

period of evolution, beings are undergoing preparation to attain their human level in the Jupiter evolution. As you know, during the Saturn, Sun and Moon evolutions—you can look it up in my book *Occult Science*[30]— the spirits of the higher hierarchies took part in the preparation of humanity. There is a description of how Angeloi, Archangeloi and Archai were involved in the development of human beings, and therefore an obvious question is whether human beings, during their Earth evolution, are perhaps involved in preparing the beings who will reach their human level during the Jupiter evolution?

This question is certainly a vital one for every sensitive person—sensitive in the sense of having been inspired by spiritual science, as we have been describing. Might it be the case that human behaviour in general during the course of Earth evolution could either help, or fail to help, the beings who may attain their human level in the Jupiter evolution? We could say, 'What can be worse than behaving in such a way during Earth evolution as to make it impossible for proper Jupiter beings to arise through our deeds?'

If we want to talk about these things we must of course take it for granted that there is a degree of good will towards this knowledge, for these are truly important secrets of initiation, the kind of initiation secrets that modern science detests as a matter of course. One certainly has to be sensitive in preparing to look at the attitude modern science is bound to have to the real truths of life.

In the previous lectures I have already tried to say a little about the way modern science necessarily relates to life. It cannot have direct access to the secrets of life. It cannot even want this; and it should not even claim to want to reach these secrets of life. It may be a good thing to make hard-boiled eggs for people who like eating eggs when they are hard-boiled, and hard-boiled eggs are useful for those who like them. But if we were to go and say that we are taking the eggs away from the hens to hard-boil them, and then let the hens hatch them out after that, we would be doing something absurd. As far as the cosmos is concerned we would be doing exactly the same thing if we were to set out to solve its secrets by means of modern science. This is exactly the same attitude as wanting to hatch hard-boiled eggs that have nothing left in them to be hatched out.

Let me use a comparison to show you how very misleading this science is that is bound up with the whole way of modern thinking, particularly when it approaches the real riddles of life. Those wanting to hold forth about whether science is helpful or harmful will usually start off by asking, 'Is science right about this or that?' If they can prove that it is right in one instance or another, they will then have absolute confidence in it as a matter of course.

Attaching so much importance to the question of whether what science says is right or not is the very attitude we have to get away from. We must reach the point of seeing that this is not the main thing when it comes to solving the riddles of life. If you see a horse-drawn cart with a man in it, you will be quite right in saying that the horses are pulling this person in the cart and drawing him along behind them. This is correct, of course; and anyone who wanted to say that the horses were not drawing the cart with the man sitting in it would obviously be wrong. But it is also true that, by guiding the horses, the man sitting in the cart is controlling the direction in which they should pull him; and that is surely the more important aspect from the point of view of the destination. Modern science can be compared to the statement of someone who denies that the man in the cart is guiding the horses, someone who will only concede that the horses are drawing the man in the cart.

If you think this comparison through in detail, you will get the right idea about the relation of modern science to modern research into truth. I have to say these things over and over again because those who base themselves on our view of the world must become increasingly capable of defending and protecting the point of view of spiritual science against the attacks of the modern world outlook. But you will be able to do this only if you have a clear idea as to the relation of modern external science to genuine research into truth. You must always approach questions of spiritual science with a certain specific attitude and a certain kind of feeling, otherwise you will not cope properly with them.

Our question concerning the beings who will reach the human level on Jupiter is indeed connected with the deepest questions of mankind's evolution on Earth. There is something in our Earth evolution that has always been a philosophical problem, namely the relation between moral behaviour and natural existence. As earthly beings we have to decide to what extent we are ruled by our instincts and are forced to obey and satisfy them, to what extent we are at the mercy of our instincts and their satisfaction because the laws of nature simply insist on their being satisfied. That is one side of human nature. In this respect we say, 'We do these things because we must. We have to eat and we have to sleep.'

But there is another sphere of human conduct on this

earth, a sphere in which we cannot say 'must', for it would lose its whole significance if we were to say 'must' here. This is the wide sphere of 'shall', a sphere where we feel that we have to follow a purely spiritual impulse rather than our instincts and everything arising out of ourselves on the natural level. 'You shall' never speaks to us from out of our instincts but directs us in a purely spiritual way. 'You shall' encompasses the sphere of our moral obligations.

There are some philosophers who cannot distinguish between what is implied by 'you shall' as opposed to 'you must'. Our present age that is almost bogged down in materialism, especially where moral life is concerned—and it will get more and more bogged down—would like to turn all the 'you shalls' into 'you musts'. We are heading for times when, in this respect, the turning of 'you shall' into 'you must' will be proclaimed with a certain amount of arrogance, and actually called psychology. Terrible considerations present themselves if we look at what has begun to develop, for example in the field of criminal psychology. It is already evident that individuals are being conceived of in such a way that people do not ask whether they have transgressed a 'you shall', but try to prove that they were driven to a destructive act out of the necessity of their nature. Strange attempts are on the increase to define crime merely as a particular case of illness. All these things arise out of a certain materialistic lack of clarity in our times regarding the distinction between 'you shall' and 'you must'.

What does this 'you shall', or in other words the categorical imperative, actually signify within the whole framework of human existence? We know that someone who obeys the 'you shall' carries out a moral action. Someone who does not obey the 'you shall' commits an immoral action. This is of course merely an everyday truth. But now let us attempt to look at 'moral' and 'immoral', not only with regard to the external maya of the physical plane but with regard to the truth and what actually exists behind physical maya. Here the moral, the ethical element corresponding to the 'you shall' appears to initiation science as something that hits you in the eye, spiritually, to put it rather crudely. If you see a person—these truths which the materialistic outlook detests have to be told sometimes—if you see someone in certain temperature and weather conditions—you notice this even better in horses, but we are not speaking about horses now—you will see him breathing out, and the breath becoming visible as vapour in the air. Obviously as far as materialistic science is concerned this breath a person exhales disperses and dissolves and has no further significance. But it has significance for someone who follows up the phenomena of life through initiation science, for he sees in the patterns of the breath the exact traces of the moral or immoral conduct of the person. A person's moral or immoral behaviour can be seen in the vaporous breath, and the breath of a person who is morally inclined is quite different from the breath of a person who is inclined to immorality. Certain more delicate qualities can only be detected in the more delicate parts of the etheric and astral aura. But the human being's moral and immoral tendencies in the ordinary sense of the words are actually visible in the etheric-astral content of the vaporous breath. The physical part of it dissolves. But what is incorporated in it does not dissolve, for it contains a demonic being which, in the case of vaporous breath, has a physical, an etheric and an astral part, only the physical is not earthy, just watery. Something with extremely differentiated forms can be seen in this exhaled breath.

When deeds arise out of love something quite different manifests compared with what comes about through deeds done out of enthusiasm, or a creative urge or the urge for perfection. But in every case the form in the exhaled breath reminds us of beings who do not exist on earth at all as yet. These beings are a preparation for the ones who will reach their human stage on Jupiter. Their forms are very changeable and will pass through further changes in the future, for these are the first advance shadowy images of the beings who will reach their human level on Jupiter.

In a sense we in our turn owe our existence to the exhalation of the Angeloi during the Moon evolution; and it is one of the moving experiences of spiritual life to know that Jupiter human beings of the future will evolve out of what we breathe out in present ages. If we approach the Bible with such knowledge in mind, and read the opening words, we can tell ourselves, 'Now we begin to understand what is meant when it says that breaths were exhaled by the Elohim and that the Elohim formed earthly man by breathing into him.'

I will confess that I would never have understood the part about the Elohim breathing the living being of man into his mouth and nostrils, if I had not known beforehand that the breath of earthly human beings also contains the first germinal beginnings of the beings who will become human on Jupiter. But Jupiter human beings can only arise from the kind of breath that owes its existence to deeds that obey the 'you shall', and which are therefore moral actions.

Thus we see how through our earthly morality we play a creative part in the whole cosmic order. Our earthly morality is indeed a creative power, and we can see that spiritual science gives us a strong impulse for moral action by telling us that we are working against the creation of Jupiter human beings if we do not act in a moral way during Earth evolution. This gives morality, the expression of 'you shall', a very real value, an existential value. Our human conduct is very strongly formed by what we acquire through spiritual science, especially as we become acquainted with real secrets regarding the cosmos.

I have already made references to similar things and mentioned at various times that speech, too, symbolizes man's own future creativity. I do not wish to dwell on this today though, but just wanted to show you what significance moral behaviour has in the whole of the cosmos.

You might now ask, 'What about immoral behaviour?' Immoral behaviour, too, comes to expression in the inner formation of the exhaled breath. But immoral behaviour imprints a demonic form in it. Demons are born through man's immoral conduct. Let us look at the difference between the demons that arise through immoral behaviour and the spiritual beings—spiritual in so far as they only achieve a watery existence on earth— the spiritual forms that are created by moral behaviour.

These beings who achieve a transitory watery existence in exhaled breath and arise from moral conduct are the kind who have an astral, an etheric and finally a physical body condensed to the level of wateriness, just as, in the

Moon evolution, we had an etheric, an astral and a physical body, and that physical body was also only condensed to a kind of wateriness. This is more or less, though not quite, what we were like during the Moon evolution. This creature arising out of moral actions and possessing a physical body, ether body and astral body has a predisposition to receive an ego, just as in the Moon evolution our physical, ether and astral bodies were predisposed to receive an ego. They are predisposed to receive an ego, and beings of this kind are qualified to undergo a regular progressive evolution in the cosmos.

The other beings, the demons created out of immoral actions, also have an astral body, an etheric body and a physical body, at the watery stage, of course, but they do not have the basis for developing an ego. They are born headless, as it were. Instead of taking up the basis for progressing along a regular evolutionary path to the Jupiter existence, they reject this basis. By doing so they condemn themselves to the fate of dropping out of evolution. But this only increases the hordes of luciferic beings, for they fall under their power. As they cannot progress in a regular way they have to become parasites. This is what happens to all the beings who reject a normal evolution; they have to attach themselves to others in order to move on. Beings who arise through immoral actions have the particular inclination to be parasites in human evolution on earth under Lucifer's leadership—to which they have succumbed—and to seize hold of the evolution of human beings before they make their physical entry into the world. They attack human beings during the embryonic stage and share their existence between conception and birth. Some of these beings, if they are strong enough, can continue to accompany the human being after birth, as seen in the phenomena of some children who are possessed.

What is brought about by the criminal demons attached as parasites to unborn children is the cause of a deterioration in the succession of the generations; it gnaws at human beings, making them less able to develop than they would be if these demons did not exist. There are various reasons for the decline of families, tribes, peoples and nations, but one of them is the existence of these criminal demon parasites during the period mentioned.

These things play an important part in Earth evolution as a whole, and they bring us into contact with deep secrets of human existence. This is often the cause of people acquiring certain prejudices and points of view even before they are born. So people are tormented by doubts and uncertainties in life, and all kinds of other things, because of these demonic parasites.

These beings cannot do very much with what human beings develop once their ego makes itself felt, but they prey on them all the more before they are born or during their earliest years. Thus we see that evil actions, too, have a significant effect in the cosmos and work creatively, but their creativity tends in the direction of the Old Moon existence. For what human beings pass through in the embryonic period, when these demonic beings can prey on them, is basically the heritage of the Old Moon evolution, which makes its appearance in all kinds of subconscious, instinctive behaviour. Something stemming from the older and better days of physical science is the instinct it preserves about the human embryonic period not being calculated according to ordinary months but lunar months; science still speaks of ten lunar months, and knows certain other things, too, concerning the connection of embryonic development with the phases of the moon.

So we see that our Earth evolution contains two tendencies. There are good deeds that contain the impulse to work creatively on Earth in preparation for Jupiter, so that man's successor on the human level can come into being. But evil deeds have also brought into our evolution the tendency to drag Earth back again to the Old Moon evolution and make it dependent on everything to do with subconscious impulses. If you think about it you will find a great number of things that are connected with these subconscious impulses; in fact, there are far more of these subconscious impulses in materialistic humanity of modern times than there were in bygone ages when people were less materialistic.

I believe that the kind of things I have been telling you about again today will make you feel what a deep impression spiritual science can make on your outlook on life, and that it will definitely not only give you theoretical knowledge but will be capable of giving human life a new direction. A time will come when life will become quite chaotic if people do not use the chance of giving it a new direction out of spiritual science. People must get beyond knowledge that is restricted to physical circumstances. Our materialistic age does not want knowledge of any other sort than the kind that is restricted to the physical body. But people must lift their knowledge out of this physical body. What we know today as the first exercises in *Knowledge of the Higher Worlds*[31] will gradually—though 'gradually' will be quite a long time—turn into something we do naturally, something we will feel to be second nature. Particularly what we call thought concentration will come naturally to people.

People will more and more feel the need to concentrate fully in their thoughts, to focus their whole soul activity on clearly defined thoughts which they hold up in front of their consciousness. While they would otherwise let their senses roam from one thing or fact to another, they will more and more often direct their thought life to definite things of their own choosing; even if only for a short time, they will concentrate on a definite thought, so as to focus their soul activity on this thought. They will then discover something that a lot of you know very well. Everyone makes a certain discovery in the course of concentrating like this. If we place a thought in the centre of our consciousness and focus all our soul activity on it, we notice the thought growing stronger and stronger. Certainly it does. But then there comes a point when it doesn't get stronger any more but gets weaker and fades away. This is an experience lots of you will have had. The thought has to fade away; it has, as it were, to die away inside. For the kind of thoughts we have at first, the way we think, arises through the instrument of the physical body, and we concentrate the kind of thinking we do by means of the instrument of the physical body until the moment when the thought dies; then we slip out of the physical body.

We would go into the unconscious altogether if, parallel with this concentration exercise, we did not attempt to do something else to maintain our consciousness even after we have slipped out of our physical body. What we have to do to maintain our consciousness when we are outside is what we call leading a calm and composed life and accepting the things of the world calmly. We can do even more than accept things with composure. We can take seriously a theory we know so well, namely, the concept of karma. What do I mean?

People are not at all inclined at first to take the idea of karma really seriously. If they have even a small mishap that hurts them, or anything at all happens to them, they sometimes get furious, but at any rate they have antipathy towards it. We encounter what we call our destiny with sympathy or antipathy. In ordinary life this has to be so; here it is essential that we feel sympathy towards some of the occurrences of destiny and antipathy towards others. To us, destiny is something that comes to meet us from outside.

If we take the idea of karma seriously, we must fully recognize our ego in our destiny and realize we ourselves are active in what happens to us through destiny; that we are the actual agents. When someone offends us it is certainly difficult to believe we are hidden inside the offender. For it may be necessary in physical life to punish the offence. But we must always keep a corner within us where we admit to ourselves, 'Even when someone offends me it is me offending myself; when someone hits me it is me hitting myself, when unpleasant blows of destiny hit me, I myself am dealing myself these blows.' We forget that we are not only within our skin but also in our destiny; we forget we are within all the so-called chance happenings of our destiny.

It is very difficult genuinely to acquire a sense of the way our own ego brings us our destiny. Yet it is true that with our own ego we bring ourselves our destiny, and we get the impulses for this in the life between death and a new birth according to our earlier incarnations, so that we can bring ourselves our destiny. We have to strive to unite with our destiny and, instead of warding off hard blows of fate with antipathy, tell ourselves more and more often, 'Through suffering this blow of destiny, that is, through meeting myself in this blow of destiny, I am making myself stronger, more vigorous and robust.' It is more difficult to form a union with your destiny in this way than to resist it, but what we lose when the thought fades can only be regained by drawing into ourselves, like this, what is outside us. We cannot stay in what is within our skin if the thought fades when we concentrate on it, but it will carry us out of ourselves if we have taken hold of our destiny, our karma, in the true sense. We thus awaken ourselves again. The thought fades, but we carry with us the identification we have grasped between our ego and our destiny, and it carries us about in the world.

This composure with respect to our destiny, this sincere acceptance of destiny, is what gives us existence when we are outside our body. Obviously this does not need to alter our life on the physical plane. We cannot always do that. But the attitude we have to acquire in a corner of our soul must be there for the moments when we really want to be able to live consciously outside our body.

Two maxims can be our guiding principles and can mean a very great deal to us. The first of these to impress upon our mind is:

Strive for the dying away of the thought in the universe.

The thought becomes a living force outside us only when it dies away in the universe. Yet we cannot unite ourselves with this living force unless we work at the content of the second maxim:

Strive for the resurrection of destiny in the 'I'.

If I achieve this, I unite the 'I' that has been reborn in the thought with the resurrected 'I' outside me.

There is a great deal in human nature, though, which makes it difficult to progress in the sense of these maxims. It is particularly hard to look at the relationship between inner and outer in the right way. The more we can learn morally from spiritual science in this connection, the better. We can learn what is moral from it in so far as certain ethical concepts can only acquire life and blood through what spiritual science can bring to them.

For instance, there are people who are perpetually complaining about other people and the awful things they do to them. They even go so far as to say that other people are persecuting them. Everything of this kind is always connected with the other pole of human nature; you only have to observe life in the right way, which means according to spiritual science as properly understood. Of course there is good reason to complain about unkindness, but in spite of this you will always find, if you go through life with vision that has been somewhat clarified by spiritual science, that most of these complaints come from egoists, and that the suspicion that everyone wants to be nasty to them arises most often in egoistic natures, whereas those with a loving disposition will not readily suspect persecution, nor that people are trying to harm them in all kinds of ways, and so on.

When such a thing is put into words like this it is easy to agree with it in theory. In fact I'm convinced that most people will admit to it theoretically, if they stop to think about it. But to live accordingly is what matters.

You may now ask, 'How do we live accordingly?' And

again the answer must be, 'You must actually live with the spiritual science you are striving for, live with it as much as you can.' That is the point. That is why spiritual science is not given in compendia or short sketches; we are trying to make spiritual science a living stream of life in which we can live and remain warm, and from which we can constantly draw stimulating impulses.

This was the main reason that led us to make the Goetheanum building a kind of focal point for this living anthroposophical work for, as I mentioned yesterday, the Goetheanum building, in its form and in the whole way it has been built, has a bodily, a soul and a spiritual nature as described by spiritual science. It is itself a token of the fact that we are striving according to impulses from the spiritual world to bring into human evolution what it so badly needs today, if the immediate future is to go the way it should. The very fact that you can perceive the being of spiritual science in the forms of the Goetheanum building makes it a sort of centre, a focal point around which the kind of anthroposophical effort can crystallize that has to become an essential part of the evolution of humanity.

We probably often say that we live in troubled times and that many things exist today which stand out in stark contrast to spiritual science. Yet our karma has allowed us to come so far as to overcome to some extent the material out of which our Goetheanum is built, so that this physical material can be a token of spiritual science even in its external form.

Each one of us can acknowledge this to him or herself, as I often do, especially in the difficult times we are going through in face of the strong attacks being made on spiritual science at present. Some people might question how much personal progress we have made, for our part, towards what ought to be crystallizing around the Goetheanum. Even if one individual or another cannot be physically present any more in the work being done on the physical plane, the fact that the building is there, and that our karma has brought it to us, is an important step forward. If we bear in mind that people with as deep an understanding of spiritual science as Christian Morgenstern, for example, remain united, even after physical death, with what the Goetheanum is intended to be, then our building can also be a token in our time of the kind of activity within our spiritual movement that is not concerned with boundaries between what is usually called life and death.

We can really feel united with this building, and therefore it can stimulate us to think the kind of serious thoughts which it is quite natural for us to think at a time like ours, when materialism is at its peak. Even if this building finds one individual or other co-operating only as a spiritual being, the building will be important for the continuation of our movement. This, you will understand, is said only to show you that our movement has a seriousness that transcends life and death.

We have encountered this seriousness to a special degree this week. And if one thing happens, why should not other things be able to happen soon, too? It is extraordinarily difficult to achieve what I have often spoken about and mentioned again this time. I have stressed that I give the strictest attention to choosing and finding exactly the right words for which I can be fully answerable to the spiritual worlds. So it is to be desired that these words shall be heard and received in the same spirit.

Times will certainly come when people will be able to be more light-hearted with regard to spiritual science. But now, right at the beginning, we must get used to taking things very, very seriously. A while ago I talked here about occult research, among other things, because I thought it would be of use to some of you. I was mainly describing the characteristics of facts. I thought these would be of help to some of you in understanding our present difficult times. But what I presented with this intention has not been treated with sufficient caution—there are one or two exceptions, of course, but I am right in saying that not everyone has treated it with the necessary discretion and regard, and it has transpired that it was repeated here and there in a form that conveyed exactly the opposite of what I said here.

I only have to think of what has been made out of something that cannot have been misunderstood because it is already in print,[32] regarding the classification of the peoples of Europe according to sentient soul, intellectual or mind soul, consciousness soul and the ego, which was truly not meant to express an order of superiority. The statements that have gone out into the world and the opposition and bad feeling they have aroused, make it easy to see that the principle of taking things quite exactly, even in such difficult cases, has not been given due consideration. If, for instance, I had ever said that the ego activity which is there among the European peoples ought to be effective in organizing the European population, I would have been talking nonsense. Yet this is one of the things that has been put about in public, where it is creating the most fearful misunderstandings and bad feelings.

I am therefore now in duty-bound not to say a word

about these things in my lectures for the time being. I must desist from any mention of them, and whereas everybody else is free to speak about whatever they like, it has become impossible for me to do so because of the way people have misconstrued what I said.

All these things go to show that we should look at ourselves and see if we cannot take the anthroposophical movement absolutely seriously. Sometimes no notice at all is taken of the sense of responsibility there is for each sentence of the communications of genuine spiritual research, when the matter is taken seriously. To stir up emotions is certainly not what spiritual science is for, nor to oppose or pacify them. So if anyone says these things are communicated in order to oppose somebody, he ought to ask himself whether the right or wrong use has been made of what has been communicated with the utmost objectivity and reverent love of truth. These are things which add even more pain to my occult experiences, which are already painful enough. Even though I cannot therefore mention certain important things, and a lot has to be omitted which I thought we could investigate soon, there are still a good many important and substantial matters left for us to discuss.

LECTURE 8

Working with Sculptural Architecture—II

Dornach, 4 January 1915

I should like to begin today by saying a few words about the boiler house that is associated with the Goetheanum, and the architectural principle underlying it. If you want to study what motivated the architectural forms of the boiler house you must bear in mind that it is part of the whole Goetheanum building and belongs to it. This fact of it belonging to the building has to be expressed in the artistic conception of the building itself, if this conception is correct. It should not be an abstraction but has to be expressed in the artistic form.

We have to look at the way the artistic forms relate to one another, and we can get closest to this if we do what people unfortunately do far too seldom, which is to think of the tremendous artistic creative activity we find if we can see the spiritual aspect of nature and recognize natural creation as a product of the spirit. I would like to draw your attention to the forms of the skeletal system because it is easiest to see there. The human skeleton, especially, is less difficult to study than any other organic forms.

You will know that I have been trying for decades to arouse some understanding in the world for the significant discoveries Goethe made in the field of anatomy and physiology, which I should like to call his second major achievement in this realm. I will not touch on the first one today but only refer to the second. This second significant discovery owes its origin to what, in the external materialistic world, one might call the combination of chance and human genius. Goethe related that one day when he was going for a walk in the Jewish cemetery in Venice he found a sheep's skull that had fallen apart at the seams.[33] Picking it up and looking at the shape of the bones the thought occurred to him, 'When I look at these skull bones, what actually are they? They are transformed vertebrae.'

You know, of course, that the spinal column which encloses the nerve strand of the spinal cord is composed of rings that fit into one another, rings with a specific shape and with bony processes. Imagine one of these rings expanding so that the hole the spinal cord passes through—for the rings fit one on top of the other— begins to get larger and the bone correspondingly thinner, stretching like an elastic material, not only horizontally but also in the other directions; then the form that arises out of this ring shape is nothing else but the bone formation that shapes our cranium. Our cranium is composed of metamorphosed vertebrae.

On the basis of spiritual science we can develop this discovery of Goethe's even further and can say today that every bone in the human body is a transformation, a metamorphosis, of a single form. The only reason we do not notice this is because we have very primitive views of what can arise through metamorphosis. If you think of a bone of the upper arm—you know of course what it looks like—a long hollow bone like that would not immediately strike you as being similar to a bone in our skull. But that is only due to our not having developed the concept of metamorphosis far enough.

The first idea you will have is that the long hollow bone has to be puffed up to create a balloon-shaped inner space, and that then you ought to arrive at the form of a cranial bone. But that is not the principle in the case of these bones. A long hollow bone would first have to be turned inside out, so you would not see the similarity it has to the cranium until you had turned it inside out like you might a glove. But when a person turns a glove inside out he expects it to resemble what it was like before, doesn't he? This is because the glove is something dead, which is quite different from something living. If the glove were alive, the following would happen when it was turned inside out. Changes would occur like the thumb and the little finger getting very long, the middle finger very short, the palm contracting, and so on. The turning inside out and the varying elasticity of the material would bring about all kinds of changes; in fact, the glove would acquire a totally different form, although it would still be the glove. But if you imagine a hollow upper arm bone being turned inside out, then a cranial bone would emerge.

You will realize that the wise powers of the Godhead in the cosmos possessed a greater wisdom than arrogant man has today, to be able to set the forces of transformation in motion that are needed to form a cranium. The intrinsic unity in nature comes from the very fact that, fundamentally, even the most dissimilar forms are transformations of one archetypal form. There is nothing in the realm of life that could have arisen except as a metamorphosis of a primary form. In the course of this metamorphosis something else happens as well. Certain parts of the primary form become larger at the expense of others, and other parts become smaller; also various limbs expand, but not all to the same extent. This produces dissimilarities, although they are all metamorphoses of the same primary form.

Look at the primary form of our whole Goetheanum building. I can only give you a very sketchy account of what I want to tell you, and only mention one point of view.

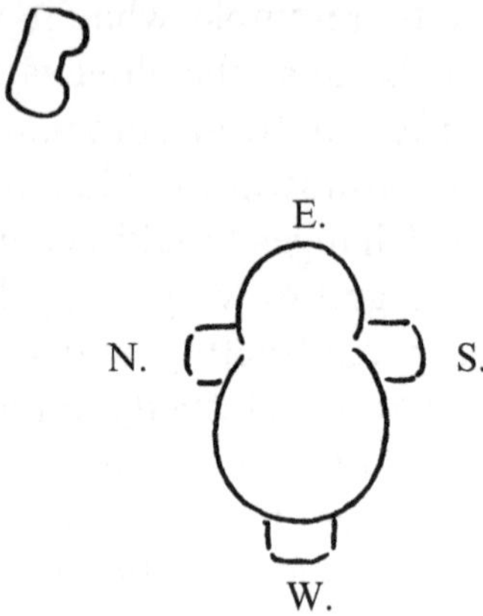

If you look at the Goetheanum you will see that it has two domes and that the domes rest on cylindrical substructures. The fact that it is a building with two domes is vital, for this double dome is an expression of the living element. If there had only been one dome then in essence our building would have been dead. The living quality of our building is expressed in the fact that, in a way, the consciousness of the one dome is reflected in that of the other, that the two domes mirror into one another, just as the part of the human being that is in the external world is reflected in the inner organs. The basic concept of the double sphere must be borne in mind in relation to anything intimately and organically connected with the Goetheanum, for anything not containing the double dome form, however hard it was to recognize, would not express the essential nature of the concept of the building. That is why this side-building must also contain the concept of the double dome.

Look at the double dome and its additional constructions. First of all we have the intersection of the two dome motifs, a feature of particular importance to which I have often referred. It represents a kind of innovation in architecture and, as you know, was worked out from the engineering point of view with the help of Herr Englert. The intersection of the two domes is of special importance in the main building because it expresses the inner connection of the two elements which mirror one another. I am giving you this concept of mirroring in an abstract way at the moment. A very great deal is contained in this intersection of the two dome motifs, an infinite amount of different aspects. The further phase of the building, the artistic phase, that expresses itself as a reflection of the concept of spiritual science, can only come into being because we have succeeded in achieving this intersection of the two dome motifs. So we have this intersection in the main building. If we were to cancel the intersection and separate the domes, we would move more towards an ahrimanic principle. If we were to bring them closer together or overlap them completely, by building one inside the other, we would approach the luciferic principle.

What we wanted to achieve was to remove the ahrimanic principle out of the main building and into an annex where the domes are pushed apart, for in the case of the annex, too, the dome concept is vital. So now imagine the domes kept apart. Imagine that on one side, this side motif [south portal of the main building] has shrunk to nothing, so the dotted line has gone, and on the other side it has grown considerably larger, and become the chimney. With the main building in mind, imagine that here [south] you have the separated domes, here is a front structure, and here the whole thing has been pushed in (a). There the whole thing has been pulled out instead of being pushed in (b), but here (a) it has shrunk to nothing instead of growing. Imagine that on the other side [the front structure of the north portal] it developed considerably, and you have the metamorphosis motif for an annex to our main building which has developed out of the primary forms. For if you imagine this getting smaller and smaller [the chimney], that coming out again and the whole thing pushed together, then the annex would be metamorphosed into the main building.

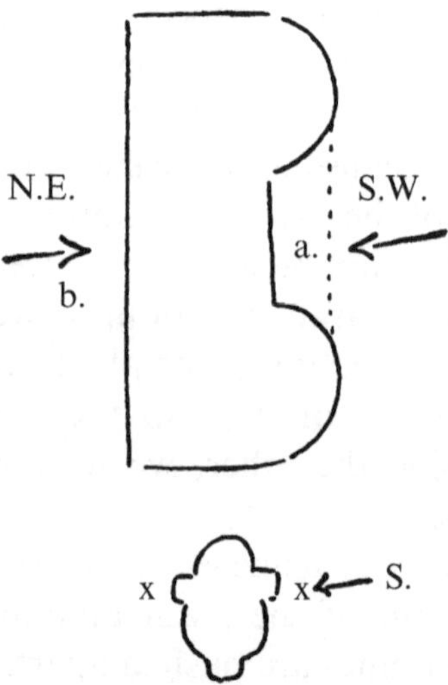

[Steiner was pointing to a model of the boiler house as he spoke.]

The point is that this metamorphosis of our main building is an adaptation to suit a specific purpose. Just as a vertebra arises out of the same primary form as a bone in the human cranium, and you can think of one changing into the other, you can also think of the main building and the annex changing from one to the other. The idea of the form can pass from one form to another if it undergoes a metamorphosis and becomes alive.

We really have to become apprentices of the creative hierarchies who created by means of metamorphosis, and learn to do the same thing ourselves.

Imagine the kind of force necessary to enlarge this insignificant-looking part on this side [north portal of the main building which becomes the chimney]. If you have a small expandable bag that you want to enlarge, you have to push it out this way and that way from inside so that it gets bigger. A force has to be there that can enlarge things and develop them. So if one of these side wings had to be puffed up, it would have to be done by a force working from inside, from here [see left-hand diagram].

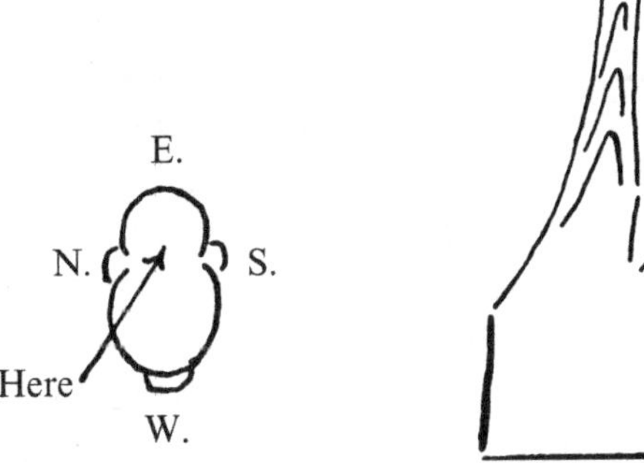

What kind of forces can they be, in there? You can study these forces in the forms of the architraves. If you imagine the forces in the architraves jumping into the side structure and pushing this up, you get this form [chimney and rear wall]. You have to try and slip inside these forms of the architraves with your formative artistic thinking and contract and expand them. Imagine that because you have slipped inside, you enlarge what is small in here. Then this form arises [chimney and rear wall]. There is no other way of going about creating things that belong together than by trying to get inside them.

This slipping into things and being inside them is another way of imitating the creative forces in nature, and it is the only way of overcoming the godforsakenness of modern industrial civilization. It would be impossible to imagine the ordinary kind of factory chimney as a product of natural creation. It only exists because there is a denial of nature's divine-spiritual forces. There is hardly anything outside in nature that you could compare with an ordinary factory chimney except possibly a rather hideous-looking asparagus plant. But that is a kind of exception. Whatever grows with the forces of earth can never go straight upwards like a factory chimney. If you want to study the forces that work in an upward direction, a tree is the best example in which to find what corresponds to the hidden forces in the earth, for a tree does not only develop a trunk in the vertical, but also has to reach out with its branches. It is obviously not the point to imitate this directly in the model; the point is to study those forces which radiate out from the earth and overcome the purely vertical direction of the tree trunk by reaching out breadthways and putting forth branches. This part here does justice to the centrifugal forces in space, in the cosmos; it does justice to what I would like to call the branching-out centrifugal forces [on the chimney].

Although I have only been able to show you the principle in the broadest outline, I could justify the principles behind this architectural form in minute detail in the case of every single surface, but it would take too long.

A form such as this is only complete when it is fulfilling its purpose. As it stands now it is not complete. It will only be complete when the heating is actually functioning inside and smoke is coming out; smoke belongs to it, it really belongs, and this has been included in the architectural form. One day when the rising of the smoke is observed clairvoyantly, and the smoke is coming out through the chimney, the spiritual part of the rising smoke will also be taken into consideration—for we shall know, when we have really observed it clairvoyantly, that the physical also contains a spiritual element. For just as you have a physical, an etheric and an astral body, the smoke also has at least an etheric part. This etheric part goes a different way from the physical: the physical part will go upwards, but the etheric part is really caught by these branchings that reach outwards. A time will come when people will see the physical part of the smoke rising while the etheric part wafts away. When this kind of thing is expressed in the form, a principle of all art is gradually being complied with, namely, the presenting of the inner essence in outer form, really making the inner essence the principle according to which the outer form is created.

As I said, I would have to do a lot of talking if I were to

go into all the details on which these architectural forms are based, which might in fact be far more interesting than those we have already discussed. One of these interesting things is that it was possible to express everything that needed to be expressed in this modern material, and build with concrete. We shall be able to go a long way with form-making in this modern material, especially in the designing of buildings in this style that are to serve modern ahrimanic civilization. In fact it is essential to do so. As I am more concerned with showing you the principle of this building and everything to do with it there is no need for me to go into any further details. This principle can be modified in many respects. For instance the dome can be modified so that it does not look like a dome any more if it is looked at solely from the geometrical-mathematical point of view and not organically, and so on. But today I wanted to discuss the particular principle of inner metamorphosis and transformation, the life principle within these. I wanted to cite this to show you in what way real artistic creativity, when it has to do with our spiritual-scientific conception, has to differ from any kind of symbolic interpretation, for that is external. It is a matter of getting an inner grasp of what you are being shown here and following the process with your whole soul.

When the building is eventually finished we do not continually want to be asked, 'What does this mean and what does that mean?', and have to witness people feeling satisfied that they have discovered the meaning of some of these things. Regarding the matter of wanting to interpret everything, we have got into some strange situations along some of the by-paths of theosophy, with respect to quite a number of poetic and literary works.

For instance, people have interpreted plays by saying that one figure signifies manas, another figure buddhi, a third figure atma, and so on. If you want to, you can of course explain everything in this manner. But we are not concerned with this kind of interpretation, but with entering into things and joining in the process of creativity that comes from the higher hierarchies and fills and forms the whole of our world. There is no need to avoid doing this just because it is more difficult than symbolic or allegorical interpretation. For it leads into the spiritual world and is the very strongest incentive for really acquiring Imagination, Inspiration and Intuition.

A vital part of the present-day impulses for change is that we acquire more and more understanding for the way the human soul rises into the realms that open up to imaginative, inspirational and intuitive observation. For these realms contain the elements that will make our world whole again, and which will lift us up out of mere maya and lead us to true reality.

It has to be stressed again and again that the new spiritual knowledge we are moving towards cannot consist of repeating the results of earlier clairvoyance. There are certainly a lot of people striving to repeat earlier clairvoyance, but the time for this clairvoyance is over, and it is only atavistic echoes of ancient clairvoyance that can possibly occur in these few individuals. The levels of human existence to which we must ascend will not open up to a repetition of ancient clairvoyance. Let us have another look at the essential basis of this new clairvoyance. We have often referred to the principle of the thing, so today let us try and approach it from another angle.

Again we will start with something we all know,

namely, that during the waking day we live with our ego and astral body in our etheric and physical bodies. But I have already emphasized during the past few days that, awake as we are between waking up and going to sleep, we are not fully awake, for something in us is still asleep. What we feel as our will is really only partly awake. Our thoughts are awake from when we wake up until we go to sleep, but using our will is something we carry out completely in a dream. This is why all the pondering about freedom of will and about freedom altogether has been in vain, because people have failed to notice that what they know about the will during the daytime is actually only a dreaming of their will impulses. On the other hand, if you have an impulse of will and make a mental image of it, then you are certainly awake. In ordinary waking life, though, we only dream with regard to how the will arises and goes over into action.

If you pick up a piece of chalk and make a mental image of picking it up, that is of course something you can have a mental image of. But with only your day consciousness and without clairvoyance you know no more about how the ego and astral body stream into your hand and how the will spreads out, than you know about a dream while you are dreaming. With ordinary waking consciousness we can only dream about the actual will, and where most things are concerned we do not even dream, we just sleep. You can clearly visualize putting a mouthful of food on your fork, and to a certain extent you can visualize chewing it, but you do not even dream about swallowing it. You are usually quite unconscious of it, as you are unconscious of your thoughts while you are asleep. So during waking life a major part of our will activity is carried out in sleep although we are awake.

If we did not sleep especially with regard to our desires and the feeling impulses bound up with them, something strange would begin to happen. We would trace the course of our actions right into our body; in everything we do out of a will impulse we would follow what goes on inside us, in our blood and throughout its whole circulation. That is, if you could follow the picking up of a piece of chalk from the point of view of the will impulse, you would follow in the whole of your blood circulation what is going on in your hand; from the inside you would see the activity of the blood and the feelings that are bound up with this. For instance, you would have an inner perception of the weight of the piece of chalk and become aware that you were seeing the nerve tracts and the flowing etheric inside them. You would experience yourself internally moving along through the activity of the blood and the nerves, which would amount to an inner enjoyment of your own blood and nerve activity. But we have to be free of this inner enjoyment of our own blood and nerve activity during earthly life, otherwise we would go through life wanting this inner pleasure to accompany everything we did. Our enjoyment of our own self would increase enormously, but as they are at present constituted human beings are not supposed to have this enjoyment. The secret of why they should not have it can be found, yet again, in a passage of the Bible, for which we ought to feel greater and greater reverence.

After the events in paradise as recorded in the paradise myth, man went on being able to eat from the Tree of Knowledge while having to stop enjoying the Tree of Life. Deriving inner pleasure in the manner described would amount to enjoying the Tree of Life, but this was not supposed to happen. I cannot develop this theme further today as it would lead too far, but through meditating on it yourselves you will be able to discover more about it.

Another thing that can have special significance for us in connection with these present lectures, and arises out of this matter, is the following: Not being able to eat from the Tree of Life means not being able to enjoy the blood and nerve activity going on within us. Yet because we know the outer world by means of our senses and our reason, something comes about that surely has something to do with this kind of inner pleasure. Whenever we perceive anything in the outer world we follow the course of our blood circulation towards the region of the senses—eyes, ears, nose and taste buds—and when we think about it we follow the nerve tracts. But we do not have the perception of what is going on in the blood circulation and nerve tracts, for what we would have perceived in the blood is reflected and mirrored by the senses, thus causing the sense impressions to arise. What is conducted through the nerve tracts is also reflected to where the nerve tracts end, where it is then mirrored as thought.

Imagine for a moment a human being who is in the following situation: Instead of merely following his blood from the outside and getting a reflection of what it does, and instead of merely following his nerves and getting a reflection, he does something which we are actually prevented from doing, which is to experience the blood and the sense nerves as they approach the eyes. This would amount to obtaining inward bodily pleasure—at least as far as those particular parts of the blood and nerves were concerned. This is what gives rise to the inner pictures of atavistic clairvoyance. What we see reflected are only pictures, something like filtered pictures of what is in the blood and the nerves. There are world secrets in the blood and nerves, but the kind of world secrets that have already been used up by creating us. It is only ourselves we get to know when we acquire Imaginations resulting from experiencing the blood flowing towards the senses, and it is only Inspirations that have the task of building up our bodily nature which we get to know when we experience ourselves in the nerves leading to our senses.

A whole inner world can arise in this manner, and this inner world can be a collection of Imaginations. When perceiving the outer physical world in a way that is proper for our Earth evolution, we perceive reflections of our blood and nerve activities. However, if we follow our blood only to the point where it flows into the senses, this amounts to indulging in inner pleasure. Then the experience of the world of Imagination is comparable to swimming in blood like a fish in water. But this world perceived in Imagination is in reality not an outside world but one that lives in our blood. If we live in the nerves leading to the senses we experience a world of Inspiration full of music of the spheres and inner pictures. Although this is cosmic, it is nothing new. It has already fulfilled its task in that it has coagulated to form our nerve and blood systems.

The kind of clairvoyance arising in this way, and leading man further into himself instead of beyond himself, is self-enjoyment, veritable self-enjoyment. This is why a kind of refined voluptuousness is brought about in people who become clairvoyant in this way and experience a world which is new to them. On the whole we must say that this kind of clairvoyance is a return to an earlier stage of evolution. For although this way of living in one's own sense organs and blood, as I have been describing to you, did not exist then in the form in which it does today, the nervous system was already there in a germinal state. This kind of perception was the normal one for human beings on Old Moon, and what they had then in the way of the beginning of nerves served to give them inner perceptions of themselves. The blood had not yet taken form in them; it was more like a warm breath coming towards them from outside, like we receive the rays of the sun. Therefore what is now, in the Earth stage, a perception of the inner blood system, was the regular, normal perception of the outside world at the Moon stage.

If this [a diagram was drawn] is the borderline between man's inner and outer world, then what are now our nerves were already there, in germinal form, in the Moon evolution. By following the course of the nerves the human being could perceive what was within him as a world of inner Imaginations. He saw that he himself belonged in the cosmos. He also had an imaginative perception of what came to him as a breath from outside and not from inside. That has now ceased, and what was outside, on Old Moon, has become internalized as the blood circulation in Earth evolution. So perception of that kind is a regression to the Old Moon evolution.

It is good to know of these things, because that kind of clairvoyance keeps on making its appearance. To acquire it does not require the hard path of meditation and concentration described in *Knowledge of the Higher Worlds*. The kind of clairvoyance that arises as a result of learning to experience one's nerves and blood as an inner enjoyment is just a more refined development of organic life, a further development of the kind of experience arising out of eating and drinking. This is certainly not mankind's present task, but is a kind of hothouse plant which arises when self-enjoyment of things such as eating and drinking is developed to a fine art. Just as a connoisseur enjoying a Rhine or a Moselle wine has an inner after-taste which is only an imagination of the taste and is not an actual taste, so some people have a subtle inner enjoyment which is their clairvoyance.

A lot of clairvoyance is nothing more than a subtle, refined, forced kind of after-taste of life. We must become aware of these things again in our time, for people have not known about these secrets nor referred to them in literature since the first half of the nineteenth century. Then came the second half of the nineteenth century, with all the discoveries that are considered so wonderful, and they certainly are wonderful from their point of view, but an understanding of those things and the finer connections of existence were lost. But what has not yet been lost—and this is said in parenthesis—is the enjoyment of the effects of the coarser things we imbibe. We continue to be able to live in the after-enjoyment of eating and drinking and, precisely in the materialistic age, have cultivated this to a considerable degree.

But humanity lives in a rhythmic cycle where things like this are concerned. Having eradicated the general feeling it used to have of indulging in self-enjoyment in the senses, nerves and blood circulation, which people had to a greater extent in the past, the materialistic age can now all the more strongly devote itself to the effects of eating and drinking. You can easily study the whole change and rapid development that has taken place in a relatively short time in this realm. You only have to compare a hotel menu of the 1870s with a present-day one [1915], and you will see the progress that has been made in the sphere of refined pleasures, in the self-enjoyment of one's own body. Yet things of this kind also move in cycles, and achievements are only carried to a certain level. Just as a pendulum can only swing to a certain point before it has to go back again, the indulgence in mere physical pleasure will also have to go in the other direction once it has reached a certain point. This will occur when the keenest epicureans, that is, people with the greatest longing for pleasures, will look at the choicest dainties and say, 'Ugh! I don't fancy that, it's just too much!' This moment will certainly come, for it is a necessary development. Everything moves in cycles.

Human beings experience the other side of life during sleep. Their thought life is asleep and it is quite natural for different conditions to prevail. As I said, it was chiefly the first half of the nineteenth century that still had insight into these matters; and the kind of clairvoyance that arises when one follows the course of one's own blood and nerve tracts was still called Pythian clairvoyance at that time, from certain memories people had, and it is indeed related to the foundations of ancient Pythian clairvoyance.[34]

Other conditions prevail during sleep. Human beings are outside their physical and etheric bodies with their ego and astral body. In ordinary life thoughts are then suppressed and devitalized. But between falling asleep and waking up they live continually with the longing for their physical body. This is precisely what sleep consists of: acquiring a longing for one's physical body from the moment one falls asleep. This rises to a climax and then urges us more and more to return to our body. When we are asleep the longing to return to our own physical body becomes stronger and stronger, and because this longing pervades the ego and the astral body like a mist, the life of thought is dimmed. Just as we cannot see the trees any more when mist closes in, we cannot experience our life of perception within us when the mist of our longing envelops us.

It can happen that this longing grows so strong during sleep that the human being does not keep it outside his physical and etheric bodies, but that his greed grows to the extent that he partly takes hold of his inner physical and ether bodies and comes into touch with the extreme ends of his blood and nerve tracts; he enters from outside, as it were, through his senses into the extreme ends of his blood circulation and his nerve tracts.

In ancient times, when man still had experiences like these with the help of the gods, it was a normal and healthy process. The old Hebrew prophets, who accomplished so much for their people, acquired their prophetic gifts through making use of the tremendous love they bore precisely for the blood and nerve composition of their own people, so that even during sleep they did not want to let go altogether of what lived physically in their people. These prophets of Jewish antiquity were seized with such longing and filled with such love that even in sleep they wanted to remain bound to the blood of the people to whom they belonged. This is precisely what gave them their prophetic gifts.

This is the physiological origin of these prophetic gifts, and splendid achievements came about through this means. This is precisely why the prophets of the various peoples had such significance for their own people, because even when they were outside the body they maintained a contact with it in this way.

As I mentioned, there was still some degree of awareness, right up to the first half of the nineteenth century, of this aspect of the life of humanity. Just as they called the other kind of clairvoyance Pythian clairvoyance, this kind of clairvoyance, which came about through contact of the ego-astral nature of man with the blood and nerve tracts of the physical body even during sleep, was called prophetic clairvoyance.

If you look at the literature of the first half of the nineteenth century you will find descriptions of Pythian and prophetic clairvoyance, even if they are not as precise as descriptions of them by the new spiritual science would be today. People now no longer know the difference between them, since they have no understanding of what they read about them in the books of the first half of the nineteenth century. But neither of these kinds of clairvoyance can really help humanity forward today. Both of them were right for olden times. Modern clairvoyance, which has to develop further and further in the future, can come about neither through enjoying what is going on in our bodies while we are awake, nor by entering into the body from outside in a sleeplike state, urged on by love—even if it is not for ourselves, but for the part of mankind to which our body belongs. Both these levels have been superseded.

Modern clairvoyance must be a third way, involving neither a taking hold of the physical body from outside, in loving longing, nor the enjoyment of the physical body from inside. Both these aspects, that which lives within and floods the body with pleasure, and that which can take hold of the body from outside, have to go out of the body if modern clairvoyance is to occur; they may only have the sort of relationship with the body, during incarnation between birth and death, in which they neither enjoy nor love the blood and nerves, either from inside or from outside, but remain connected with the body only while freely abstaining from such self-enjoyment or self-love. Some connection with the body has to be maintained, of course, otherwise it would mean dying. The human being must remain bound to the body that belongs to him in physical incarnation on earth, and this must be done by means of other parts which are to some extent remote, relatively remote from the activity of blood and nerves. Detachment from the activity of blood and nerves must be achieved.

When a person no longer indulges in enjoyment of self inwardly along the tracts leading to the senses, nor pushes from outside as far in as the senses, but has a kind of relationship with himself, both from inside and from outside, in which he can actually take living hold of what symbolizes the death of physical life, then the proper condition has been reached. For we actually die physiologically because we are capable of developing our skeletal system. When we are capable of taking hold of the skeleton, which folk wisdom recognizes as the symbol of death, and which is as remote from the blood system as it is from the nervous system, then we come to what we can call clairvoyance out of spiritual science, which is more advanced than either Pythian or prophetic clairvoyance.

With the clairvoyance of spiritual science we take hold of the whole and not just part of the human being, and it actually makes no difference whether we take hold of it from inside or outside, for this kind of clairvoyance can no longer be an act of enjoyment. Instead of being a subtle enjoyment it is an opening up and rising into the divine-spiritual forces of the All. It is a uniting with the world. It is no longer an experiencing of the human being and the mysteries that have been woven into him, but an experiencing of the deeds of the beings of the higher hierarchies, whereby one truly lifts oneself out of self-enjoyment and self-love. Man must become like a thought, an organ of the higher hierarchies, just as our thoughts are organs of our soul. To be thought, pictured and perceived by the higher hierarchies, that is the principle of spiritual-scientific clairvoyance; to be taken up, not to take up oneself.

I would like to express the wish that what I have just been saying might become a real object for your meditation, for precisely what I have been telling you today can bring many, many things to life in all of you and enliven the actual impulses of our movement to an ever greater extent. How seriously we have to take the enlivening of our spiritual-scientific movement has often been spoken of during our days together here. We would be able to bring to realization a further part—I will not say of what was intended—but of what ought to be intended within this movement of spiritual science, if as many people as possible would resolve to think about these three kinds of knowledge of higher worlds in a living way, so that our thoughts become clearer and clearer about what, at bottom, we all intend, and which can become so easily confused with something that can be had much more comfortably.

Truly, it is not for nothing that we work on from one lecture cycle to the next to accumulate more and more concepts and ideas. Studying these concepts and ideas is not needless work, for it is precisely the way to prepare the soul impulses that will lead us to real clairvoyance in the sense of spiritual science. But by only dipping here and there into the ideas given by anthroposophy you can sometimes make a chink in one or another part of the human organization, causing fragments of Pythian and prophetic clairvoyance to arise, enough to make people proud of themselves. In such cases we often hear statements like, 'I don't need to study everything in detail, and I don't need what the lecture cycles say. I know all that, I knew it already,' and so on. Many people are still satisfied with the principle of living in a few Imaginations which we could call blood and nerve Imaginations. A lot of people fancy they possess something really special if they have a few blood and nerve Imaginations. But this is not what can lead us to selfless co-operation in mankind's development. Indulging in blood and nerve Imaginations actually leads to a heightening of self-enjoyment, to a more subtle form of egoism, in consequence of which the cultivation of spiritual science can be the very thing that breeds an even more subtle kind of egoism than you ever find in the outside world.

It is taken for granted that one is never referring to the present company nor to the Anthroposophical Society, which is, of course, here. But it should be permissible to mention that there are societies in which some people manage to turn the principles in their favour, and without really giving their unselfish support, make use of one thing or another, preferably those things which stimulate blood or nerve Imaginations, and then imagine they can spare themselves the rest. As a result they acquire an atavistic clairvoyance, or perhaps not even that, but only the feelings that accompany that kind of imaginative clairvoyance. These feelings are not an overcoming of egoism, just a heightened form of it. You find in such societies—the Anthroposophical Society excepted for politeness' sake—that although it would be people's duty to develop love and harmony out of the depths of their hearts from one member to another, there is instead disharmony, quarrelsomeness, people telling tales about one another, and so on. I can say things like this for, as I said, I always make an exception of the members of the Anthroposophical Society.

This shows us that dark shadows are thrown precisely where a strong light is about to appear. I am not finding such faults because I imagine these things can be exterminated overnight. That cannot happen, because they occur quite naturally. But at least each person can work on himself; and it is not a good thing if people are not made aware of these things.

It is thoroughly understandable that precisely because a particular movement has to be founded, the shadow sides also make themselves felt in such societies, and that what is rampant in outside life is far more rampant within them. Yet it always gives one a bitter feeling if this happens in societies which, by their very nature—otherwise there would be no point in having a society—ought to develop some degree of brotherliness, a certain loyalty; yet precisely because people come closer together, certain faults that are short-lived in the outside world develop all the more fiercely. Since present company, the Anthroposophical Society, is excepted, it will give us all the more opportunity to think and reflect about these things quite objectively and impartially, so that we really know what they are about. Then if we come across them somewhere, we shall know them for what they are and not imagine, if someone thinks he has a particularly deep grasp of Anthroposophy, that we cannot understand why faults which occur in the outside world appear much more strongly in that person. We shall understand it, but we shall also know that we have to combat it. Sometimes we cannot combat things until we have really understood them.

This is another example showing how closely connected life is with spiritual science and that the outlook of spiritual science cannot, in fact, achieve its aim unless it becomes an attitude to life, an art of life, and is brought into the whole of life. How wonderful it would be if—let us now say in the Anthroposophical Society—all the various human relationships proved to be as harmonious as we are trying to make the forms of our Goetheanum building, where they change from one to the next and each is in harmony with the other; if it could be the same in life as it is in the Goetheanum, and the whole life of our society could be like the wonderful co-operation among the people engaged in building the Goetheanum, so that even this very building activity is a harmonious and noble image of what comes to expression in the building itself.

Thus, the inner significance of the life principle of our Goetheanum building and the inner significance of the co-operation among the souls—well, perhaps I would rather not say that ... The inner significance of the harmony of the forms of our building ought to find its way into all the various human relationships in our society, and its inner formative force ought to stand before us as a kind of ideal. I should just like to assure you that the wrong words did not slip out by mistake just now, when I stopped in mid-sentence. I stopped quite deliberately, for sometimes the thing is said without actually saying it.

To summarize the theme I have given in many variations over the past few days: What I want to recommend to you most warmly is not only to look at the thoughts and ideas of spiritual science, the results of spiritual research, with your intellect and reason, but to take what lives in spiritual science into your hearts. For the salvation of mankind's future progress really depends on this. This can be said without presumption, for anyone can see it if they try at all to study the impulses of our evolution and the signs of our times. With these thoughts I will close the series of lectures I have ventured to give you at the turn of the year.

Notes and References

1 R. Steiner, 'How does one bring reality of being into the world of ideas?' (in GA 156). Typescript available at The Library, Rudolf Steiner House, London.

2 For the cultural eras see R. Steiner, *Occult Science: An Outline* (GA 13), tr. G. & M. Adams (London, Rudolf Steiner Press, 1984).

3 R. Steiner, *Secrets of the Threshold* (GA 147), tr. R. Pusch (New York, Anthroposophic Press, 1987).

4 R. Steiner, *Occult Reading and Occult Hearing* (in GA 156), tr. D. Osmond (London, Rudolf Steiner Press, 1975), lecture of 3 October 1914.

5 The Council of Clermont, convoked by Pope Urban II in 1095, which launched the Crusade movement.

6 The foundation stone of the first Goetheanum was laid on 20 September 1913. The building was destroyed by arson during the night of 31 December 1922 to 1 January 1923.

7 R. Steiner, *Theosophy* (GA 9), tr. M. Cotterell, A.P.Shepherd (London, Rudolf Steiner Press, 1970).

8 The first verse of a ninth-century hymn to John the Baptist. Guido of Arezzo (990–1050) used the initial syllable of each line to name the first six ascending notes of the C major scale.

9 See R. Steiner, *Theosophy*, op. cit., Chapter 9.

10 R. Steiner, *Knowledge of the Higher Worlds* (GA 10), tr. D. Osmond & C. Davy (London, Rudolf Steiner Press, 1985).

11 In other contexts the seven soul experiences on the path of initiation have been termed by Rudolf Steiner 'the seven great secrets of life'. For more detail see R. Steiner, *Zur*

Geschichte und aus den Inhalten der ersten Abteilung der Esoterischen Schule 1904 bis 1914 (GA 264) (Dornach, 1984), p. 248f.

12 R. Steiner, *A Road to Self Knowledge* and *The Threshold of the Spiritual World* (GA 16/17), tr. H. Collison & M. Cotterell (London, Rudolf Steiner Press, 1975).

13 See the lecture 'Technology and Art' in this volume.

14 The first Goetheanum was built from wood, while concrete was used for the boiler house.

15 Joseph Englert (dates unknown), chief engineer during the building of the first Goetheanum.

16 'The Portal of Initiation' in R. Steiner, *Four Mystery Dramas* (GA 14), tr. A. Bittleston (London, Rudolf Steiner Press, 1982).

17 See *Norske Folkviser*, ed. T. Lammers (Kristiana, Aschehoug & Co., 1910).

In addition to the lecture in the present volume (31 December 1914) Rudolf Steiner had spoken about the Norwegian *Dream Song of Olaf Åsteson* on 1 January 1912 (English typescript Z229) and 7 January 1913 (English typescript S17); on each occasion there had been a recitation of the Song by Marie Steiner. In German all three lectures or addresses are in the volume *Der Zusammenhang des Menschen mit der elementarischen Welt* (GA 158), (Dornach, 1993). It was Ingeborg Möller-Lindholm, the Norwegian poetess (1878–1964), who drew Rudolf Steiner's attention to the legend, and it is largely due to her initiative that this extraordinary folk epic has acquired such an important place in the anthroposophical movement. She kindly permitted the use of the record she made of her conversations with Rudolf Steiner. Several points from a lecture she gave on the Dream Song are also included in these Notes with her kind permission, and are attributed accordingly.

Notes on the Dream Song by Ingeborg Möller-Lindholm of Lillehammer

In June 1910 Dr Steiner gave a cycle of lectures in Oslo entitled *The Mission of the Individual Folk Souls in Relation to Teutonic Mythology* (GA 121), (London, Rudolf Steiner Press, 1970). As I lived in Oslo and had a large room at my disposal, I invited to tea about 40 anthroposophical friends who had come to town for this occasion. Dr Steiner and Frau Marie Steiner had also agreed to come. I asked Dr Steiner the previous day whether he could tell us something about the unusual Norwegian folk epic 'The Dream Song of Olaf Åsteson'. He smiled amiably and said he would first have to have read or heard it. I saw the point of this. Then he himself made the suggestion that he should arrive the next day an hour before the other guests, so that I could read the song to him and make a rough translation. And that is what happened.

While I read it to him, Dr Steiner sat with his eyes closed, listening intently. He was obviously deeply affected by the remarkable content of the song. After tea the Dream Song was read aloud in Norwegian by a member of the Society, whereupon Dr Steiner gave a short but moving lecture on the Song. In particular he dwelt on the fact that events in the Song had taken place during the time of the 12 holy nights when extra-terrestrial influences are at their weakest. He also gave special mention to the name of Olaf Åsteson. Olaf or Oleifr means 'the one left behind' after his predecessors have gone. He is the only one who passes on the blood of the father of the generations. Åst means love; so he is the 'Son of Love'.

Dr Steiner asked me to translate the song into German. He himself did not know Norwegian, let alone the old dialect in which the Dream Song had been written down, which today is difficult even for Norwegians. To begin with I made the excuse that I did not have a sufficient command of the German language to convey the wonderful musical rhythm. Dr Steiner said that did not matter, I should just translate the song literally word for word, so that he could get a more accurate picture of the content. I did this in the course of the autumn and sent him the translation, which was very prosaic and in many respects extremely inadequate. Later on Rudolf Steiner put the song into its own characteristic rhythms and gave several lectures on it. It was also used for eurythmy performances, especially at Christmas time.

Dr Steiner told me in 1913 that I should not think that Olaf the Saint was the original Olaf Åsteson. (St Olaf, a Norwegian king, died in 1035 at the battle of Stiklestad, championing the cause of Christendom.) There had been several people with the name of 'Olaf Åsteson', said Dr Steiner. It was a kind of Mystery title.

Dr Steiner was in Norway again after the First World War, in 1921 and 1923. On these occasions he stayed with engineer Ingerö. Mrs Ragnild Ingerö, who died a few years ago, told me that Dr Steiner had talked with her about the Dream Song. He had meanwhile gone into it further and discovered new things. One of these was that the song was much older than people believed. It originated around AD 400. At that time there had been a great Christian initiate in this country. He had founded a Mystery school in southern Norway; the place was not named. His Mystery name had been Olaf Åsteson, and the song described his initiation. Originally, so Dr Steiner said, the song had been much longer and had twelve sections, one for each sign of the zodiac. The song described Olaf Åsteson's journey through the whole zodiac and what he saw and experienced there. Today we only have fragments of the original song. The aforesaid Mystery school continued into the early Middle Ages. The leader was always called Olaf Åsteson.

Dr Steiner said that in course of time he would publish these facts and other important things connected with the Song. However, he did not want to do this until he had found certain external proofs of his findings. He thought he would be able to find these. But the burning of the Goetheanum, excessive work and finally illness and death prevented this intention being realized. Now these indications are all we have.

I have given much thought to these findings of Dr Steiner's and have come to the conclusion that this Mystery school was possibly Skiringssal. This place is or rather was in Vestfold, a region in south-western Norway. In old legends it has always been described as a holy place. Vikings who died on foreign soil wished to be buried in Skiringssal. There was also a *kaupang* (a market) there. Archaeologists are excavating things at present which they assume to be remains of this market place. Up to now, though, nobody has been able to prove conclusively where Skiringssal is. At the time of the Mystery school it was on the coast; however, loam deposits have meanwhile 'pushed' the place further inland. Skiringssal means 'The Hall of Purification'. *Skirn* means baptism or purification (old Norwegian).

Where did the first Olaf Åsteson come from? It has been historically proved that Irish-Scottish monks were in this country long before Christianity was officially introduced. According to legend, Joseph of Arimathea came to the British Isles as early as the first century, and began his missionary work there. There have been Mystery centres in Ireland since very early times. The tribes on the neighbouring islands were heathen. The Irish-Scottish

Church, also called the Culdee Church, arose as a result of the confluence of the work of the Christian missionaries and ancient Druid wisdom. It flourished in many places as early as AD 300 and 400. There were churches, schools and monasteries, despite the fact that these were always under attack from powerful heathen tribes of the neighbourhood. Many priests and monks died a martyr's death. This Culdee Church was based in particular on the Gospel of John and the preaching of the Apostle John. It was like the early communities of Christians and contrasted strongly with the Petrine or Roman Catholic Church. But the latter was victorious. The Culdee Church was destroyed and dissolved in the year AD 664. It sent a lot of missionaries to various European countries both before and after being externally destroyed. This Church was definitely of an esoteric nature. Many things suggest that the first Olaf Åsteson was a representative of this spiritual stream.

'Among these Norwegian people, who still possess many things in their popular tongue that approach very closely the threshold of occult secrets, possibilities existed for souls to remain connected longer with everything living and working behind outer material phenomena,' said Rudolf Steiner in his lecture on Olaf Åsteson in Berlin on 7 January 1913. (I.M.)
Pauline Wehrle rendered the poem into English from the German version.

18   The Gjallar Bridge spans the mystical River Gjöll, which separates the realms of the spiritual world. (I.M.)

19   Olaf goes on to tell how he journeyed across the zodiac. It is particularly obvious here that a large part of the song is missing, as Landstad (a well-known Norwegian psalm writer) also says in his commentaries. The only constellation that is mentioned is the Dog (*Canis major*), although this constellation is outside the zodiac, as is the

Snake (*Serpens*). But the Bull (*Taurus*) is a sign of the zodiac that has significance for him. After he has journeyed through the zodiac, Olaf is prompted to take a different path and goes along the Milky Way (*Vintergaten*). There is an old belief that the Milky Way leads to Paradise, the realm of the blessed. (I.M.)

20 'Brooksvalin' is a strange old word that Landstrad translates as 'the forecourt of oppression'. It follows from the song that Olaf now returns to the zodiac and goes into the sign of the Scales. (I.M.)

21 Grutte Graubart—Ahriman. (I.M.)

22 Wherever Christianity had spread there were pictures of Michael holding the scales in one hand. In the other he often had a lance or sword with which he pierced the dragon. He is presented like this in innumerable church paintings and sculptures, as also on the north portal of the Cathedral Church in Trondheim. At this point the epic part of the song is virtually over, though a few verses follow in which Olaf preaches to his fellow men in the manner of the Holy Scriptures: 'They shall rest from their labours, yet their deeds will follow them.' Landstad was told that the song had been used for the death watch in the past. The song was to help the soul at the start of its journey in the other world. (I.M.)

23 Hugo Bergmann, Ph.D. (1883–1975) went to school with Franz Kafka and subsequently studied philosophy in Prague and Berlin. The philosopher Marti and his teacher Franz Brentano on the one hand and Martin Buber and Rudolf Steiner on the other were formative influences in his life. In 1920 he went to Jerusalem where he became Rector of the university in 1935. For the centenary celebration of Rudolf Steiner's birth in 1961 he gave a lecture before the Philosophical Society of the Hebrew University (see *Die Drei* 1962/1). When he died he was still working on a 5-volume 'History of Modern Philosophy'

in which he emphasized Goethe's importance. (See *Die Drei* 1984/10.)

24 R. Steiner, *The Threshold of the Spiritual World*, op. cit.

25 R. Steiner, *The Spiritual Guidance of Man and Humanity* (GA 15), tr. H. Monges (New York, Anthroposophic Press, 1983).

26 'The Portal of Initiation', Scene 8, in R. Steiner, *Four Mystery Dramas*, op. cit.

27 It has not been possible to discover who this was.

28 'The Souls' Awakening' in R. Steiner, *Four Mystery Dramas*, op. cit.

29 This is presumed to be Hermann Bahr (1863–1934), whose 'active mind' Rudolf Steiner often mentioned. In his *Kritik der Gegenwart* (Augsburg 1922), Bahr admitted, '...if I reveal what my vocation was: in 1898 I could have succeeded Burckhard as director of the *Burgtheater...*' It is known that this position was offered him in 1918 and that he declined, giving as the reason that 'his' *Burgtheater* had long since 'fallen to pieces'. Whether an earlier offer was made in 1913 round about his fiftieth birthday is not known.

30 R. Steiner, *Occult Science*, op. cit.

31 R. Steiner, *Knowledge of the Higher Worlds*, op. cit.

32 R. Steiner, *The Mission of the Individual Folk Souls*, op. cit.

33 In the annals of 1790 and in the essay 'Bedeutendes Fördernis durch ein einziges geistreiches Wort' (1822) (One clever word leads to an important step forward).

34 The Pythia was always the designation of the Apollo priestess at Delphi. By various means she worked herself into a state of ecstasy in order to impart the oracle. For more details on the ancient clairvoyance of prophets and sibyls, see the lecture of 29 December 1913 in R. Steiner, *Christ and the Spiritual World and The Search for the Holy Grail* (GA 149), tr. C. Davy & D. Osmond (London, Rudolf Steiner Press, 1983).

Publisher's Note Regarding Rudolf Steiner's Lectures

The lectures contained in this volume have been translated from the German which is based on stenographic and other recorded texts that were in most cases never seen or revised by the lecturer. Hence, due to human errors in hearing and transcription, they may contain mistakes and faulty passages. Every effort has been made to ensure that this is not the case. Some of the lectures were given to audiences more familiar with anthroposophy; these are the so-called 'private' or 'members' lectures. Other lectures, like the written works, were intended for the general public. The differences between these, as Rudolf Steiner indicates in his *Autobiography*, is twofold. On the one hand, the members' lectures take for granted a background in and commitment to anthroposophy; in the public lectures this was not the case. At the same time, the members' lectures address the concerns and dilemmas of the members, while the public work speaks directly out of Steiner's own understanding of universal needs. Nevertheless, as Rudolf Steiner stresses: 'Nothing was ever said that was not solely the result of my direct experience of the growing content of anthroposophy. There was never any question of concessions to the prejudices and preferences of the members. Whoever reads these privately printed lectures can take them to represent anthroposophy in the fullest sense. Thus it was possible without hesitation—when the complaints in this direction became too persistent—to depart from the custom of circulating this material "for members only". But it must be borne in mind that faulty passages do occur in these reports not revised by myself.' Earlier in the same chapter, he states: 'Had I been able to correct them [the private lectures] the restriction [for members only] would have been unnecessary from the beginning.'